Lecture Notes in Mathematics

Edited by A. Dold and B. Eckmann

591

G. A. Anderson

Surgery with Coefficients

Springer-Verlag
Berlin · Heidelberg · New York 1977

Author

Gerald A. Anderson
Department of Mathematics
Pennsylvania State University
University Park
PA 16802/USA

AMS Subject Classifications (1970): 57B10, 57C10, 57D65

ISBN 3-540-08250-6 Springer-Verlag Berlin · Heidelberg · New York
ISBN 0-387-08250-6 Springer-Verlag New York · Heidelberg · Berlin

Printed in Germany

Printing and binding: Beltz Offsetdruck, Hemsbach/Bergstr.
2141/3140-543210

INTRODUCTION

This set of notes is derived from a seminar given at the University of Michigan in 1973, and portions of the author's doctoral thesis. It is intended to give a reasonably complete and self-contained account of surgery theory modulo a set of primes.

The first three chapters contain the background material necessary to describe the theory. Chapter 1 is mainly definitions and notation and contains no new ideas, with the exception of relative localization and colocalization of spaces. Included is a sketch of the immersion classification theorem of Hirsch and Haefliger-Poenaru.

Chapter 2 contains the theory of local Whitehead torsion. The definition differs from the one given by Cappell and Shaneson, but is justified by a Whitehead-type local collapse-expansion theorem. Chapter 3 discusses the theory of spaces which satisfy Poincare duality with coefficients in a ring, including the construction of a local Spivak normal fibration. Normal invariants modulo a set of primes are described and the homotopy groups of the classifying space G_P/H are computed.

Chapter 4 contains the main surgery obstruction theorem. Briefly, groups are constructed to measure the obstruction to finding a homotopy equivalence (over a ring and with given torsion) cobordant to a given map. Below the middle dimension, the technique is due to Milnor and Wallace. Considering homotopy equivalences over the integers, the simply connected case is essentially done by Kervaire and Milnor, and globalized by Browder and Novikov; the general case is due to Wall. We show that the obstruction lies in a Wall group of a localized group ring.

Surgery over a field was first considered by Petrie and Passman, and Miscenko noticed that Wall's groups behaved nicely away from the prime 2. More recently, Connelly, Giest and Pardon have considered rational surgery (in the non-simple case), and the methods of Cappell and Shaneson (which uses rings with a local epimorphism $\mathbb{Z}\pi \to R$) also apply. The general case, with rings of the form $R\pi$, is due to the author in his thesis.

Chapter 5 gives the geometric definition of surgery groups, and the generalization to manifold n-ads. Quinn's approach is also briefly discussed. Finally, the periodicity theorem, in the non-simple case, is proved.

Chapter 6 describes the result of changing rings in surgery groups by means of a long exact sequence. Corollaries include a Rothenberg-type sequence, the general periodicity isomorphism and determination of the kernel of $L^s_{2k-1}(\mathbb{Z}\pi) \to L^s_{2k-1}(\mathbb{Q}\pi)$, π finite, by simple linking forms, generalizing the original odd-dimensional surgery obstructions due to Wall and clarified by Connelly.

Finally, five appendicies are included: Whitehead torsion notions for n-ads, the algebraic construction of $L_n(\pi \to \pi';R)$, the computation of $L_n(\mathbb{Z}^k;R)$, surgery on embedded manifolds, and homotopy and homology spheres. The reference has been arranged into categories. Undoubtedly, some errors and omissions have occurred in this arrangement, but I hope the general drift is helpful to the reader.

A number of people have been of great help in writing these notes. I am indebted to my thesis advisor C.N. Lee for many helpful suggestions and discussions. I would also like to thank Dennis Barden, Allan Edmonds, Steve Ferry, and Steve Wilson, who participated in the seminar, Frank Raymond, Jack Mac Laughlin and W. Holstztynski.

Massachusetts Institute of Technology

TABLE OF CONTENTS

Chapter 1. Preliminaries 1

1.1 Modules 1
1.2 Homology and Cohomology with Twisted Coefficients 2
1.3 Δ-Sets 4
1.4 Microbundles, Block Bundles and Spherical Fibrations 7
1.5 The Immersion Classification Theorem 11
1.6 Intersection Numbers 14
1.7 Algebraic K-Theory 19
1.8 Localization 23

Chapter 2. Whitehead Torsion 28

Chapter 3. Poincare Complexes 39

3.1 Poincare Duality 39
3.2 Spherical Fibrations and Normal Maps 45

Chapter 4 Surgery with Coefficients 54

4.1 Surgery 54
4.2 The Problem of Surgery with Coefficients 57
4.3 Surgery Obstruction Groups 60
4.4 The Simply Connected Case 74
4.5 The Exact Sequence of Surgery 80

Chapter 5. Relative Surgery 82

5.1 Handle Subtraction and Applications 82
5.2 Geometric Definition of Surgery Groups 85
5.3 Classifying Spaces for Surgery 94
5.4 The Periodicity Theorem, Part I 96

Chapter 6. Relations Between Surgery Theories 101

6.1 The Long Exact Sequence of Surgery with Coefficients 101
6.2 The Rothenberg Sequence 105
6.3 The Periodicity Theorem, Part II 109
6.4 Simple Linking Numbers 110

Appendix A. Torsion for n-ads 122

Appendix B. Higher L-Theories 124

Appendix C. L Groups of Free Abelian Groups 127

Appendix D. Ambient Surgery and Surgery Leaving a Submanifold Fixed 129

Appendix E: Homotopy and Homology Spheres.................. 135

References.. 138

Symbol Index.. 154

Index... 156

Chapter 1. Preliminaries

1.1. Modules.

Let Λ be a ring (not necessarily commutative) with involution, i.e. a map $\Lambda \to \Lambda$, written $\lambda \mapsto \lambda^*$, so that

(a) $(\lambda_1+\lambda_2)^* = \lambda_1^* + \lambda_2^*$

(b) $(\lambda_1\lambda_2)^* = \lambda_2^*\lambda_1^*$

(c) $\lambda^{**} = \lambda$.

We will usually assume $1 \in \Lambda$. $\Lambda^{\cdot}$ denotes the group of units in Λ. Unless otherwise stated, all Λ-modules will be finitely generated and right. Let M be a Λ-module. Then M inherits a left Λ-module structure by defining $\lambda \cdot m = m \cdot \lambda^*$.

The <u>dual of M</u> is defined by $M^* = \mathrm{Hom}_\Lambda(M,\Lambda)$ with Λ-module structure given by $(f\cdot\lambda)(m) = \lambda^* f(m)$, $f \in M^*$, $\lambda \in \Lambda$. If M and N are Λ-modules, we define $M \otimes_\Lambda N$ by giving N a left Λ-module structure as above. In the case $N=\Lambda$, $M \otimes_\Lambda \Lambda$ is a Λ-module with $(x \otimes \lambda)\mu = x \otimes \mu^*\lambda$.

A Λ-module is <u>free</u> if it is isomorphic to a direct sum of copies of Λ. M is <u>projective</u> if there is a Λ-module N so that $M \oplus N$ is free. M is <u>stably free (s-free)</u> if we may choose N to be free, i.e. $M \oplus \Lambda^k$ is free for some k. If M is s-free, a <u>stable basis (s-basis)</u> is a basis for some $M \oplus \Lambda^k$.

The main example of Λ will be a group ring, $\Lambda = R\pi$ for some (usually commutative) ring R with 1, π a multiplicative group with a homomorphism $w:\pi \to \{\pm 1\}$. The ring $R\pi$ is defined to be the set of all finite sums $\sum n_g \cdot g$, $n_g \in R$, $g \in \pi$. The involution is given by $(\sum n_g \cdot g)^* = \sum w(g) n_g g^{-1}$.

1.2. Homology and Cohomology with Twisted Coefficients.

Let X be a finite CW-complex, $\pi = \pi_1(X)$ and $w:\pi \to \{\pm 1\}$ a homomorphism. Let $\Lambda = \mathbb{Z}\pi$ and M a Λ-module. Define

$$H_i(X;M) = H_i(C_*(X) \otimes_\Lambda M)$$

$$H^i(X;M) = H_i(\mathrm{Hom}_\Lambda(C_*(X),M)),$$

where $C_*(X)$ is the chain complex of cellular chains in the universal cover $\tilde{X}$; $C_*(X)$ is a chain complex of free and based Λ-modules. If $\tilde{X}$ is not compact, we use cohomology with compact supports. We write $H_i(X)$, $H^i(X)$ for $H_i(X;\Lambda)$, $H^i(X;\Lambda)$.

We can define this alternately as follows: w determines an element in $H^1(X;\mathbb{Z}/2\mathbb{Z})$ and so a double cover $E \to X$. Let $\mathbb{Z}/2\mathbb{Z}$ act on $\mathbb{Z}$ non-trivially and define $\mathbb{Z}^t$ to be the bundle associated to E with fiber $\mathbb{Z}$. Now $\tilde{X} \to X$ is a principal π-bundle and so define $\mathcal{M}$ to be the associated bundle with fiber M. Let $\mathcal{M}^t = \mathcal{M} \otimes \mathbb{Z}^t$. Then

$$H_i(X;M) = H_i(X;\mathcal{M}^t)$$
$$H^i(X;M) = H^i(X;\mathcal{M})$$

where we use bundle (or sheaf) homology and cohomology.

If σ is an n-cell in $\tilde{X}$, cap product defines

$$\sigma\cap: C^q(X) \to C_{n-q}(X) \otimes \Lambda \cong C_{n-q}(X)$$

where $C^q(X) = \mathrm{Hom}_\Lambda(C_q(X),\Lambda)$. This extends linearly to chains in $C_n(X)$ and, in fact, to infinite chains since we are using compact supports.

This defines $\xi\cap: H^q(X) \to H_{n-q}(X)$ for $\xi \in H_n(X)$. If M is a Λ-module, define $\xi\cap: H^q(X;M) \to H_{n-q}(X;M)$ by the composition

$$\mathrm{Hom}_\Lambda(C_q(X),M) \cong C^q(X) \otimes_\Lambda M \to C_{n-q}(X) \otimes_\Lambda M.$$

If $f: X \to Y$ is a map, $f_\#: \pi_1(X) \cong \pi_1(Y)$, then define

$$K_i(X;M) = \ker(f_*: H_i(X;M) \to H_i(Y;M))$$
$$K^i(X;M) = \mathrm{coker}\,(f^*: H^i(Y;M) \to H^i(X;M)).$$

The condition $\pi_1(X) \cong \pi_1(Y)$ isn't necessary, but will suffice for our purposes.

A map $f: X \to Y$ with $f_\#: \pi_1(X) \cong \pi_1(Y)$ is a <u>homology equivalence over R</u> if $f_*: H_*(X;R\pi) \to H_*(Y;R\pi)$ is an isomorphism. X and Y have the same R-<u>homology type</u> if there is a sequence $X = Z_0, Z_1, \ldots, Z_m = Y$ and homology equivalences over R, $Z_i \to Z_{i+1}$ or $Z_{i+1} \to Z_i$, for each i.

1.3. Δ-Sets.

Let Δ be the category with objects Δ^n, the standard n-simplex, $n=0,1,\ldots$, and morphisms generated by the face maps ∂_i^n. A Δ-set is a contravariant functor from Δ to the category of sets. We define Δ-groups, etc., similarly. A Δ-map between Δ-sets is a natural transformation between the functors.

If X is a Δ-set, then $X(\Delta^k)$ is called the set of k-simplices of X. An ordered simplicial complex K has associated to it a Δ-set $D(K)$ defined by $D(K)(\Delta^k)$ = the set of k-simplices of K.

We can also define an inverse to D, i.e. a functor S from Δ-sets to topological spaces so that $S(D(K))$ is homeomorphic to K, for K a simplicial complex. To do this, let X be a Δ-set and form the disjoint union $\overline{X} = \bigcup_{n=0}^{\infty} X(\Delta^n) \times \Delta^n$, where $X(\Delta^n)$ has the discrete topology and we regard $\Delta^n = \{(t_0,\ldots,t_{n+1}) \in \mathbb{R}^{n+2} \mid 0=t_0 \leq t_1 \leq \ldots \leq t_{n+1} = 1\}$.

The maps $\partial_i^n : \Delta^{n-1} \to \Delta^n$ and $\delta_i^{n+1} : \Delta^{n+1} \to \Delta^n$ are then defined by $\partial_i^n(t_0,\ldots,t_n) = (t_0,\ldots,t_i,t_i,\ldots,t_n)$ and $\delta_i^{n+1}(t_0,\ldots,t_{n+2}) = (t_0,\ldots,\hat{t}_i,\ldots,t_{n+2})$.

Then we let $S(X) = \overline{X}/\sim$, where $\sim$ is the equivalence relation defined by $(\delta_i^n x_n, a_{n-1}) \sim (x_n, \partial_i^n a_{n-1})$, $(\partial_i^{n+1} x_n, a_{n+1}) \sim (x_n, \delta_i^{n+1} a_{n+1})$ for $x_n \in X(\Delta^n)$, $a_{n-1} \in \Delta^{n-1}$, $a_{n+1} \in \Delta^{n+1}$, $i=0,\ldots,n$.

S(X) is called the <u>geometric realization of X</u>. See Milnor [A11], Gabriel, Zisman [A6] for a complete discussion.

A Δ-set X is <u>Kan</u> if any Δ-map $(\overline{\Delta^n - \partial_i{}^n \Delta^{n-1}}) \to X$ admits an extension to Δ^n. The property of being Kan is important and we describe now a process for converting a Δ-set into a homotopically equivalent Kan Δ-set (Kan [A9]). If X and Y are Δ-sets, let $\Delta(X,Y)$ be the Δ-set defined by $\Delta(X,Y)(\Delta^k)$ = the set of Δ-maps $X \times \Delta^k \to Y$. Define $Ex^1(X) = \Delta(*,X)$, $Ex^k(X) = Ex^1(Ex^{k-1}(X))$, $Ex^\infty(X) = \varinjlim Ex^k(X)$.

Then $Ex^\infty(X)$ is Kan and has the same homotopy type as X.

In general, any notion used for simplicial complexes can be used for Δ-sets. This is expounded fully in Rourke and Sanderson [B10]. We repeat some definitions here.

Let H be a Kan Δ-group. A <u>principal H-bundle</u> over X is an orbit map $\pi: E \to X$, where E has a free H-action. If $K_H(X)$ denotes the set of isomorphsim classes of principal H-bundles over X, then in [B10] there is constructed a classifying space BH and a natural equivalence of $K_H(\ )$ and [,BH], the set of homotopy classes of maps to BH.

More generally, if A is a Kan Δ-monoid, then we define a <u>principal A-fibration</u> to be a Δ-map of pointed Δ-sets $\pi: E \to X$ so that

(a) π is a Kan fibration (i.e. satisfies the lifting property with respect to the pair $(\Delta^n, \overline{\dot{\Delta}^n - \partial_i{}^n\Delta^{n-1}}))$,

(b) $\pi^{-1}(*) = A$,

(c) there is an action of A on E, $E \times A \xrightarrow{\mu} E$, so that

$$\begin{array}{ccc} E \times A & \xrightarrow{\mu} & E \\ \text{proj.} \downarrow & & \downarrow \pi \\ E & \xrightarrow[\pi]{} & X \end{array}$$

commutes.

Again there is a classifying space BA and a natural equivalence of $h_A(X)$, the set of homotopy classes of principal A-fibrations over X, and $[X,BA]$.

As usual, if H is a Δ-group and A is a Δ-monoid with H a submonoid of A, then the fiber of the map $BH \to BA$ is ${}^{A}/H$.

We now define the Δ-groups and Δ-monoids needed to study bundles. Let H be one of TOP, PL, or DIFF.

Define H_q to be the Δ-set such that $H_q(\Delta^k) =$ the set of zero and fiber preserving H-homeomorphisms $\sigma:\Delta^k \times \mathbb{R}^q \to \Delta^k \times \mathbb{R}^q$. This means $\sigma|\Delta^k \times 0$ is the identity and σ commutes with projection to $\mathbb{R}^q$. Define $\tilde{H}_q$ by $\tilde{H}_q(\Delta^k) =$ the set of zero and block preserving H-homeomorphisms

$\sigma: \Delta^k \times I^q \to \Delta^k \times I^q$ (i.e. $\sigma(K \times I^q) = K \times I^q$ for each subcomplex $K \subset \Delta^k$).

Let R be a ring and define $G_q(R)$ by $G_q(R)(\Delta^k) =$ the set of zero and fiber preserving homology equivalences over R of pairs $\sigma: (\Delta^k \times \mathbb{R}^q, \Delta^k \times 0) \to (\Delta^k \times \mathbb{R}^q, \Delta^k \times 0)$ (i.e. $\sigma^{-1}(\Delta^k \times 0) = \Delta^k \times 0$). Define $\tilde{G}_q(R)$ similarly but with block preserving instead of fiber preserving.

Define $H = \varinjlim H_q$, $\tilde{H} = \varinjlim \tilde{H}_q$. We have H_q, $\tilde{H}_q$, H, $\tilde{H}$ are Δ-groups, and $G_q(R)$, $\tilde{G}_q(R)$ are Δ-monoids. Also, $TOP \simeq \widetilde{TOP}$, $PL \simeq \widetilde{PL}$, $DIFF \simeq \widetilde{DIFF}$, and $G_q(R) \simeq \tilde{G}_q(R)$ for all q. According to Milnor [B8], $DIFF_q \simeq GL(q,\mathbb{R}) \simeq O_q$.

1.4. Microbundles, Block Bundles and Spherical Fibrations.

In the previous section, we defined principal bundles with structure groups H_q or $\tilde{H}_q$, or $G_q(R)$. In this section we define geometrically bundles associated to these and relate them to the Δ-set definitions. As before, let $H = TOP$, PL or $DIFF$.

Definition. (Rourke, Sanderson [B9]). Let K be a simplicial complex. An H-block bundle over K, written ξ^q/K, is a space $E(\xi) \supset K$ so that

(1) if $\sigma \in K$ is an n-cell, then there exists an (n+q)-ball $B_\sigma \subset E(\xi)$ so that $(B_\sigma, \sigma) \overset{H}{\cong} (I^{n+q}, I^n)$.

(2) $E(\xi) = \bigcup_{\sigma \in K} B_\sigma$.

(3) $\mathrm{Int}(B_{\sigma_1}) \cap \mathrm{Int}(B_{\sigma_2}) = \emptyset$ if $\sigma_1 \neq \sigma_2$.

(4) $B_{\sigma_1} \cap B_{\sigma_2} = \bigcup_{\sigma \in \sigma_1 \cap \sigma_2} B_\sigma$.

The trivial block bundle ε^q/K is defined by $E(\varepsilon^q) = K \times I^q$. If ξ^q/K is a block bundle and $L \subset K$ is a subcomplex, then the restriction $\xi^q|L$ over L is defined by $E(\xi^q|L) = \bigcup_{\sigma \in L} B_\sigma$.

If ξ^q/K and η^p/L are block bundles, then define the product block bundle, $\xi \times \eta/K \times L$, by $E(\xi \times \eta) = E(\xi) \times E(\eta)$. The Whitney sum of ξ and η over K is defined by $\xi \oplus \eta = \xi \times \eta|\Delta(K)$ where $\Delta(K) \subset K \times K$ is the diagonal.

If ξ^q/K is a block bundle and $f:L \to K$, then the induced bundle, $f^*\xi^q/L$, is defined to be $L \times E(\xi^q)|G(f)$ where $G(f) = \{(x,y) \in L \times K | f(x) = y\}$ is identified with L.

Let ξ^q and η^q be block bundles over K. A bundle isomorphism is an H-homeomorphism $f:E(\xi) \to E(\eta)$ so that $f|K = 1_K$ and $f(B_\sigma(\xi)) = B_\sigma(\eta)$ for all $\sigma \in K$. A map $\sigma \times I^q \to E(\xi|\sigma) \subset E(\xi)$, $\sigma \in K$, is a chart for ξ if it is a bundle isomorphism of ε^q/σ and $\xi^q|\sigma$. A maximum collection of charts is called an atlas.

Theorem 1. ([B9]) If M is a H-manifold and N a regular neighborhood of M, then N is a H-block bundle over M.

Thus if M is a H-manifold, we can define the tangent block bundle of M, $\tilde{T}M$, to be a regular neighborhood of $\Delta(M)$ in $M \times M$. (see [B11] for the case $H = TOP$. There are some dimensional considerations in this case).

We now show how to associate a principal $\tilde{H}_q$-bundle to an H-block bundle. Let ξ^q/K be a block bundle and let $X = D(K)$. We define $D(\xi)$ to be the bundle $E \to X$ where

$$E(\Delta^k) = \{h|h:\Delta^k \times I^q \to E(\xi|\sigma) \text{ is a chart, } \sigma \in K, \text{ where } h|\Delta^k \times 0:\Delta^k \times 0 \to \sigma \text{ is the identity}\},$$

$E(\Delta^k) \to X(\Delta^k)$ sends $h:\Delta^k \times I^q \to E(\xi|\sigma)$ to σ, and the action $E(\Delta^k) \times \tilde{H}_q(\Delta^k) \to E(\Delta^k)$ is defined by composition. (in our notation, $\tilde{H}_q(\Delta^k)$ is the set of block preserving self H-isomorphisms of $\varepsilon^q|\Delta^k$).

Conversely, if ξ^q is a principal $\tilde{H}_q$-bundle $\pi:E \to D(K)$, we define an H-block bundle $S(\xi)/K$. To do this, choose a section $s:K \to S(E)$. If σ is an n-simplex of K, then $\partial_i s(\sigma) = s(\partial_i\sigma)\cdot F_{i,\sigma}$ for some unique $F \in \tilde{H}_q(\Delta^{n-1})$. Define $S(\xi)$ by $E(S(\xi)) = \bigcup_{\sigma\in K} \varepsilon^q|\sigma$, where we identify $\varepsilon^q|\partial_i\sigma$ and $(\varepsilon^q|\sigma)|\partial_i\sigma$ by the isomorphism given by $F_{i,\sigma}$ and the ordering of vertices.

These constructions give the following:

Theorem 2. There is a 1-1 correspondence between isomorphism classes of H_q-block bundles over K and principal H q-block bundles over $D(K)$.

Corollary 1. The functor which associates to K the set of isomorphism classes of H q-block bundles over K is naturally equivalent to $[K,B\tilde{H}_q]$.

Definition. (Milnor [B8]). An H q-microbundle over K is a diagram $K \overset{i}{\to} E \overset{j}{\to} K$, where i,j are H-maps, so that $j \circ i = 1_K$ and for $x \in K$ there exist neighborhoods U of x in K and V of i(x) in E and an H homeomorphism $h: V \to U \times \mathbb{R}^q$ so that

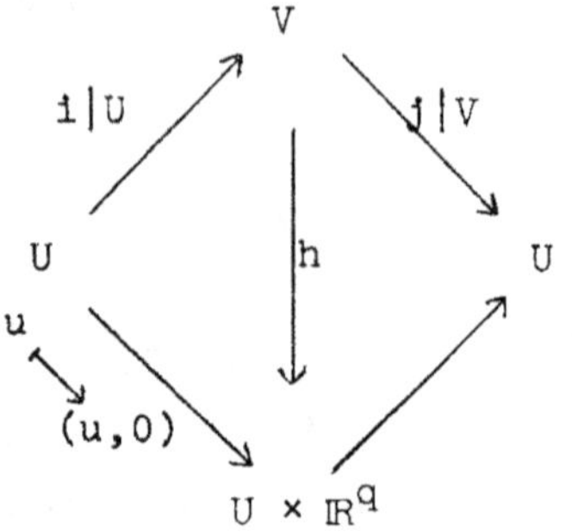

commutes.

The tangent microbundle of an H-manifold M, TM, is defined by $M \overset{\Delta}{\to} M \times M \to M$,

In the same manner of proving Theorem 2, we can prove the following theorem.

Theorem 3. The functor which associates to K the set of isomorphism classes of H q-microbundles over K is naturally equivalent to $[K, BH_q]$.

Definition. (see Spivak [E19] for the case $R=\mathbb{Z}$). Let K be as above. An R q-spherical fibration over K is a fibration $\pi: E \to K$ with fiber F so that $\pi_1(F) = 0$ and F has the R-homology type of S^q.

We define Whitney sum, induced bundles, etc., as usual.

Theorem 4. The functor which associates to K the set of fiber homotopy equivalence classes of R q-spherical fibrations over K is naturally equivalent to $[K, BG_q(R)]$.

1.5. The Immersion Classification Theorem.

In this section, manifolds may be TOP, PL, or DIFF. We use the microbundle definition of tangent bundle and assume the reader is familar with elementary properties of immersions. Submanifolds will always be assumed to be locally flat in the TOP and PL cases.

Let M be a compact submanifold of $\mathbb{R}^n$ and N a manifold with $\dim N = n$. Define Δ-sets $\mathrm{Imm}(M,N)$ and $R(TM,TN)$ by

$\mathrm{Imm}(M,N)(\Delta^k)$ = the set of germs of immersions $f: \Delta^k \times U \to \Delta^k \times N$, commuting with projection, where U is an open neighborhood of M in $\mathbb{R}^n$, and we identify f and f' if they agree on $\Delta^k \times (U \cap U')$,

$R(TM,TN)(\Delta^k)$ = the set of germs of bundle monomorphisms $F: \Delta^k \times TU \to \Delta^k \times TN$, commuting with projection, U an open neighborhood of M in $\mathbb{R}^u$, so that the map $\Delta^k \times TU \to \Delta^k \times U \times N$ given by

$(t,u,u') \to (t,u,\pi F(t,u,u'))$,

$t \in \Delta^k$, $(u,u') \in TU$, is an immersion,

where $\pi : TN \to N$ is the bundle

projection.

Choose an immersion $f : \Delta^k \times U \to \Delta^k \times N$, $M \subset U \subset \mathbb{R}^n$, and define $df : \Delta^k \times TU \to \Delta^k \times TN$ by $df(x,u,u') = (x, f_x u, f_x u')$ where $f(x,u) = (x, f_x u)$. This defines a differential map $d : \mathrm{Imm}(M,N) \to R(TM,TN)$.

The following theorem, due to Hirsch [B4] in the smooth case, Haefliger and Poenaru [B3] in the PL case and Lees [F2] in the topological case, is an important tool in surgery theory.

<u>Immersion Classification Theorem</u>. <u>If M has a handle body decomposition with all handles of index $<n$, then d is a homotopy equivalence</u>.

We will give a sketch of the proof, omitting the messy details.

<u>Lemma 1</u>. <u>Let $k < n$. Then the maps</u>

$$\Phi : \mathrm{Imm}(D^k \times D^{m-k}, N) \to \mathrm{Imm}(S^{k-1} \times D^{m-k}, N)$$

$$\Psi : R(T(D^k \times D^{m-k}), TN) \to R(T(S^{k-1} \times D^{m-k}), TN)$$

<u>defined by restriction are fibrations</u>.

Proof: This is just a matter of verifying the homotopy lifting property. See Lees [F2].

Lemma 2. The theorem is true for $M=D^n$.

Proof: See [B2] or [B3].

Let $\phi \in \mathrm{Imm}(S^{k-1} \times D^{m-k})$ and let $\mathrm{Imm}_\phi(D^k \times D^{m-k},N)$ be the fiber of Φ over ϕ and $R_\phi(T(D^k \times D^{m-k}),TN)$ the fiber of Ψ over $d\phi$.

Lemma 3. $d:\mathrm{Imm}_\phi(D^k \times D^{m-k},N) \to R_\phi(T(D^k \times D^{m-k}),TN)$ is a homotopy equivalence for $k < n$.

Proof: By induction on k, using Lemma 2.

Proof of Theorem:

The proof is by induction on handles in M. Suppose $M = M_0 \cup D^k \times D^{m-k}$, $M_0 \cap (D^k \times D^{m-k}) = S^{k-1} \times D^{m-k}$, and that the theorem is true for M_0. Since M is locally flat in $\mathbb{R}^n$, $D^k \times D^{m-k}$ is contained in a coordinate neighborhood of R^n. Thus by Lemma 1, $\mathrm{Imm}(M,N) \to \mathrm{Imm}(M_0,N)$ and $R(TM,TN) \to R(TM_0,TN)$ are fibrations, and the fiber over $f \in \mathrm{Imm}(M_0,N)$ ($df \in R(TM_0,TN)$) is $\mathrm{Imm}_\phi(D^k \times D^{m-k},N)$ ($R_\phi(T(D^k \times D^{m-k}),TN)$), where ϕ is the image of f under the map $\mathrm{Imm}(M_0,N) \to \mathrm{Imm}(S^{k-1} \times D^{m-k},N)$.

We have a commutative diagram

$$\begin{array}{ccc}
\mathrm{Imm}_\phi(D^k \times D^{m-k},N) & \longrightarrow & R_\phi(T(D^k \times D^{m-k}),TN) \\
\downarrow & & \downarrow \\
\mathrm{Imm}(M,N) & \longrightarrow & R(TM,TN) \\
\downarrow & & \downarrow \\
\mathrm{Imm}(M_0,N) & \longrightarrow & R(TM_0,TN)
\end{array}$$

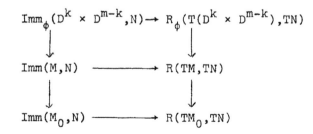

where the vertical maps are fibrations.

Since by Lemma 3 and induction, the top and bottom maps are homotopy equivalences, the middle map is also.

1.6. Intersection Numbers.

Suppose N^n is a manifold and V^v, W^w are simply connected submanifolds intersecting transversally in a finite set of points, $v+w=n$. Let $* \in N$ be a basepoint and choose a local orientation at $*$. Choose paths p_v, p_w from $*$ to basepoints $*_v \in V$, $*_w \in W$. For $x \in V \cap W$, define g_x to be the element $p_v \alpha \beta p_w^{-1}$ in $\pi_1(N,*)$ where α is a path in V from $*_v$ to x and β is a path in W from x to $*_w$, missing all other intersections. Let $\varepsilon_x = \pm 1$ according to whether or not the loop g_x preserves the local orientation at $*$.

Define the <u>intersection number</u> of V and W by

$$\lambda(V,W) = \sum_{x \in V \cap W} \varepsilon_x g_x \in \mathbb{Z}\pi, \quad \pi = \pi_1(N,*).$$

We can define $\lambda(V,W)$ if V and W are immersed submanifolds in the same manner.

To give an algebraic interpretation of λ, we need to define the <u>homology intersection pairing</u>. We assume N is closed for simplicity. For $u_1 \in H_v(N)$, $u_2 \in H_w(N)$, define $u_1 \cdot u_2 \in \mathbb{Z}\pi$ to be $\langle u'_1 \cup u'_2, [N] \rangle \in H_n(N) = \mathbb{Z}\pi$,

where $u_1' \in H^{n-v}(N)$, $u_2' \in H^{n-w}(N)$ correspond to u_1, u_2 under Poincare duality.

Theorem 1. Let $f:V^v \to N$, $g:W^w \to N$ be immersions. Then

$$\lambda(V,W) = f_*[V]\cdot g_*[W].$$

Proof: Let R,S be regular neighborhoods of V,W in N, and $T(R) \in H^v(R,R-f(V))$, $T(S) \in H^w(S,S-g(V))$ the Thom classes. Let $T'(R)$ be the image of $T(R)$ under the map

$$H^v(R,R-f(V)) \cong H^v(N,N-f(V)) \xrightarrow{j^*} H^v(N)$$

and $T'(S)$ similar. Then
$$\begin{aligned} f_*[V] &= f_*([R]\cap T(R)) \\ &= [N]\cap j^*T(R) \\ &= [N]\cap T'(R). \end{aligned}$$

Thus $f_*[V]\cdot g_*[W] = \langle T'(R)\cup T'(S),[N]\rangle$.

Let $i:N \to (N,N-f(V)\cap g(W))$. Then $\langle T'(R)\cup T'(S),[N]\rangle = \langle T(R)\cup T(S),i_*[N]\rangle$. Also if $[N]_x \in H_n(N,N-x)$ is a local orientation, then $i_*[N] = \sum_x (\phi_x)_*[N]_x$, where $\phi_x:(N,N-f(V)\cap g(W)) \to (N,N-x)$.

It is clear that

$$\begin{aligned} \varepsilon_x g_x &= \langle r^*T(R)\cup s^*T(S),[N]_x\rangle \\ &\quad \text{where } r:(N,N-x) \to (N,N-f(V)),\ s:(N,N-x)\to(N,N-g(W)) \\ &= \langle T(R)\cup T(S),(\phi_x)_*[N]_x\rangle. \end{aligned}$$

Thus
$$\begin{aligned} \sum \varepsilon_x g_x &= \sum\langle T(R)\cup T(S),(\phi_x)_*[N]_x\rangle \\ &= f_*[V]\cdot g_*[W]. \end{aligned}$$

If $V^v \subset N$ is an immersed simply connected submanifold we can define self-intersection numbers similarly. Note we must have $2v=n$.

By general position, we can assume the self-intersections of V in N are transverse double points, i.e. locally a self-intersection is the transverse intersection of two branches of V.

Let $x \in V$ be a self-intersection point. We can compute the intersection number $\varepsilon_x g_x$ as above, taking one branch arbitrarily first. If we compute the intersection number taking the other branch first, the answer will be $((-1)^v w(g_x)\varepsilon_x)g_x^{-1}$, where $w:\pi_1(N) \to \{\pm 1\}$ is the orientation homomorphism.

Thus the self-intersection number $\mu(V) = \sum \varepsilon_x g_x$ is well-defined in $\mathbb{Z}\pi/I_v$ where I_v is the ideal of $\mathbb{Z}\pi$ generated by elements of the form $a - (-1)^v w(a)a^{-1} = a - (-1)^v a^*$, $a \in \pi$.

Theorem 2 (Whitney Lemma). Let N,V,W be as above with $v,w \geq 3$. Assume $x,y \in V \cap W$ satisfy $\varepsilon_x g_x = -\varepsilon_y g_y$. Then there is an isotopy of N taking V to V', transverse to W, so that $V' \cap W = V \cap W - \{x,y\}$.

Proof: Let P,Q be paths through V,W, joining x and y. Since $\varepsilon_x g_x = -\varepsilon_y g_y$, the loop PQ^{-1} is trivial in N. Thus there is a map $f:D^2 \to N$ so that $f(S^1)$ is the loop PQ^{-1}. Since $\dim N \geq 6$, we may assume f is an embedding.

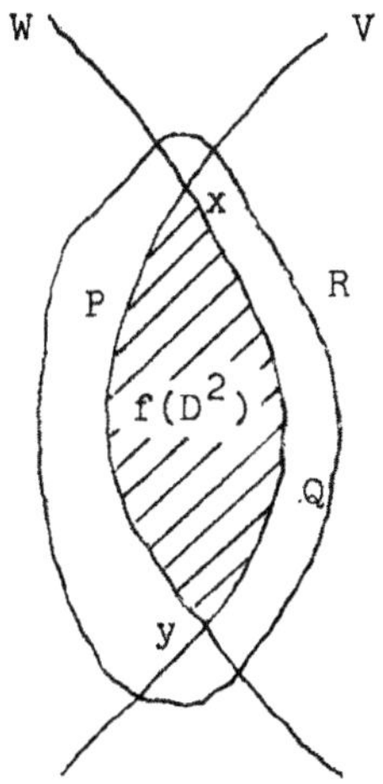

Let R be a regular neighborhood of $f(D^2)$ in N. Use this to construct an isotopy which pulls V to the other side of W. For details see Milnor [D19], Theorem 6.6 or Rourke & Sanderson [B12], Theorem 5.12.

<u>Corollary 1</u>. <u>(1) Suppose $f:V^v \to N^n$ and $g:W^w \to N$ are immersions which intersect transversally in a finite set of points</u>, $v,w \geq 3$, <u>and $\lambda(V,W) = 0$. Then f is regularly homotopic to $f':V \to N$ so that $f'(V) \cap g(W) = \phi$</u>.

(2) <u>If $f:V^v \to N^n$</u>, $2v = n \geq 6$, <u>is an immersion with $\mu(V) = 0$, then f is regularly homotopic to an embedding</u>.

There is a relative version of Whitney's Lemma for manifolds with boundary, which we leave to the reader. However, the following corollary will be useful. The proof is similar to Theorem 2.

Corollary 2. *Suppose* N^n *is a manifold,* $n \geq 6$, *with the inclusion inducing* $\pi_1(\partial N) \cong \pi_1(N)$. *Let* $f:(V^v,\partial V)\to(N,\partial N)$, $g:(W^w,\partial W) \to (N,\partial N)$ *be immersions which intersect transversally,* $v,w \geq 3$, $V,W,\partial V,\partial W$ *simply connected. Then* f *is regularly homotopic to* $f':(V,\partial V) \to (N,\partial N)$ *so that* $f'(V)\cap g(W) = \phi$.

Theorem 3. *Let* $f,g:S^k \to M^{2k+1}$, *be disjoint embeddings and* $x \in \mathbb{Z}\pi_1(M)$. *Then there are regular homotopies* $F,G:S^k \times I \to M \times I$ *from* f,g *to embeddings* f',g' *so that* $\lambda(F,G) = x$.

Proof: Since intersections are additive, we can assume $x = \pm\alpha$, $\alpha \in \pi_1(M)$. Join $f(S^k)$ and $g(S^k)$ to the basepoint of M by paths P_1 and P_2. We join $f(S^k)$ to $g(S^k)$ by the path $P_2^{-1}\alpha P_1$. Let F be a regular homotopy which deform a small disc $\subset f(S^k)$ along this path and across a disc transverse to $g(S^k)$. Let $G = g \times 1$. If the sign of x is -1, then reverse orientation in the disc.

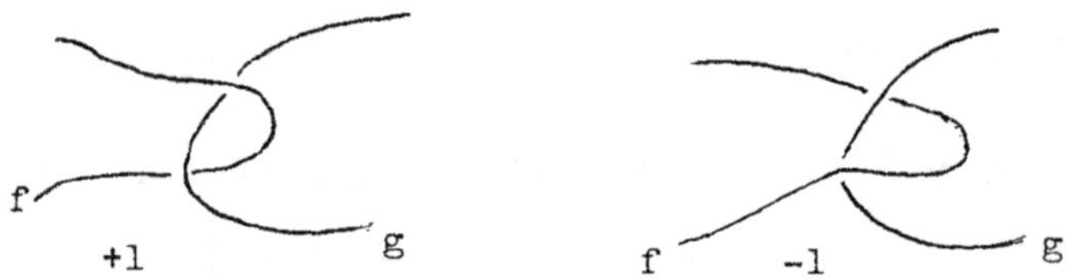

It also follows that we can find regular homotopies with arbitrary self-intersection number.

1.7. Algebraic K-Theory.

Let Λ be a ring. We describe here some functors which are both useful in geometry and will serve as a model for L-theory in surgery. A good reference is Milnor [C8].

Definition. Let $K_0(\Lambda)$ be the abelian group with generators $[M]$, where M is a projective Λ-module, and $[M]$ is its isomorphism class, with relations $[M] + [N] = [M \oplus N]$. We define a ring structure on $K_0(\Lambda)$ by $[M]\cdot[N] = [M \otimes N]$.

Examples: (1) If Λ is a principal ideal domain, then $K_0(\Lambda) \cong \mathbb{Z}$, generated by $[\Lambda]$.

(2) If $\Lambda = \mathbb{R}^X$, the ring of continuous functions $X \to \mathbb{R}$, X compact, then $K_0(\Lambda) \cong K(X)$, the topological K-functor. The correspondence is given by $\xi \mapsto \Gamma(\xi)$, sending a vector bundle ξ to the $\mathbb{R}^X$-module of continuous sections $\Gamma(\xi)$.

There is a natural splitting $K_0(\Lambda) = \tilde{K}_0(\Lambda) \oplus \mathbb{Z}$ defined by rank, and $\tilde{K}_0(\Lambda)$ is the set of isomorphism classes of projective modules $[M]$, with the equivalence relation $[M] \sim [N]$ if $M \oplus \Lambda^r \cong N \oplus \Lambda^s$ for some r,s.

The reduced K-functor $\tilde{K}_0(\mathbb{Z}\pi)$ measures the obstruction to finiteness for a CW-complex X dominated by a finite CW-complex, $\pi = \pi_1(X)$, Wall [C10], and the obstruction to finding a boundary for an open manifold M, Siebenmann [C9].

Let $GL(n,\Lambda)$ be the group of invertible $n\times n$ matrices over Λ. We regard $GL(n,\Lambda) \hookrightarrow GL(n+1,\Lambda)$ by $A \mapsto \begin{pmatrix} A & 0 \\ 0 & 1 \end{pmatrix}$. Let $GL(\Lambda) = \bigcup_n GL(n,\Lambda)$, the infinite general linear group.

Let E_{ij}, $i \neq j$, denote the matrix with 1 in the (i,j)-th spot and zeroes elsewhere. The matrix $e_{ij}^{\lambda} = I + \lambda E_{ij}$, $\lambda \in \Lambda$, is called <u>elementary</u>. The subgroup of $GL(n,\Lambda)$ generated by elementary matrices is denoted $E(n,\Lambda)$. Let $E(\Lambda) = \bigcup_n E(n,\Lambda)$.

<u>Lemma. (Whitehead). $E(\Lambda) = [GL(\Lambda),GL(\Lambda)]$.</u>

Proof: We have $e_{ij}^{\lambda} = e_{ik}^{\lambda} e_{kj}^{1} e_{ik}^{-\lambda} e_{kj}^{-1}$, so every elementary matrix is a commutator. Conversely, if $A,B \in GL(n,\Lambda)$, then in $GL(2n,\Lambda)$,

$$\begin{pmatrix} ABA^{-1}B^{-1} & 0 \\ 0 & I \end{pmatrix} = \begin{pmatrix} A & 0 \\ 0 & A^{-1} \end{pmatrix} \begin{pmatrix} B & 0 \\ 0 & B^{-1} \end{pmatrix} \begin{pmatrix} (BA)^{-1} & 0 \\ 0 & BA \end{pmatrix}.$$

Given any matrix X, we can write

$$\begin{pmatrix} X & 0 \\ 0 & X^{-1} \end{pmatrix} = \begin{pmatrix} I & X \\ 0 & I \end{pmatrix} \begin{pmatrix} I & 0 \\ I-X^{-1} & I \end{pmatrix} \begin{pmatrix} I & -I \\ 0 & I \end{pmatrix} \begin{pmatrix} I & 0 \\ I-X & I \end{pmatrix}$$

and $\begin{pmatrix} I & X \\ 0 & I \end{pmatrix}$ is in $E(2n,\Lambda)$. Thus $E(\Lambda) = [GL(\Lambda),GL(\Lambda)]$.

Define the Whitehead group of Λ by $K_1(\Lambda) = {}^{GL(\Lambda)}/E(\Lambda)$.

More categorically, we can define $K_1(\Lambda)$ to be the abelian group with generators (M,f), M a free Λ-module, $f:M \to M$ a Λ-isomorphism, with relations

$$(M,f) = (N,g) \text{ if there is a } \Lambda\text{-isomorphism } h:M \to N \text{ with } hf = gh$$

$$(M,f) + (N,g) = (M \oplus N,\ f \oplus g)$$

$$(M,f) + (M,g) = (M,fg).$$

See Bass [B2] or Swan [B11].

Definition. A free and based Λ-module is a pair (M,B), where M is a free Λ-module and $B = \{b_1,\ldots,b_n\}$ is an ordered basis for M. Another basis $B' = \{b_1',\ldots,b_n'\}$ for M is preferred if the map $M \to M$, $b_i \mapsto b_i'$, is elementary.

If Λ is commutative, let $SL(\Lambda)$ be the matrices in $GL(\Lambda)$ with determinant 1 and define $SK_1(\Lambda) = {}^{SL(\Lambda)}/E(\Lambda)$. Then $K_1(\Lambda) = \Lambda^{\bullet} \oplus SK_1(\Lambda)$.

Define $\overline{K}_1(\Lambda) = K_1(\Lambda)/\{\pm 1\}$, where $\{\pm 1\} \subset \Lambda^{\bullet} = GL(1,\Lambda)$.

Examples: (1) If Λ is $\mathbb{Z}$ or a field, then $SK_1(\Lambda) = 0$.

(2) If $\Lambda = \mathbb{Z}\pi$, π an elementary abelian p-group of rank k, p an odd prime, then $SK_1(\Lambda)$ is an elementary abelian p-group of rank ${}^{p^k-1}/p-1 - \binom{p+k-1}{p}$.

If $\mathbb{Z} \subsetneq R \subset \mathbb{Q}$, then $SK_1(R\pi) = 0$ for π as above. See [C1].

(3) If Λ is a skew field, then $K_1(\Lambda) = {}^{\Lambda^{\bullet}}/[\Lambda^{\bullet},\Lambda^{\bullet}]$ and $GL(n,\Lambda) \to K_1(\Lambda)$ is the determinant.

The main application of K_1 is to torsion. Let $C: C_n \xrightarrow{\partial} C_{n-1} \to \cdots \xrightarrow{\partial} C_0$ be a chain complex with C_i a free and based Λ-module. Suppose C is acyclic, $H_*(C) = 0$. Then by Spanier [A15], Theorem 4.2.5, there exists $\delta: C_i \to C_{i+1}$ so that $\partial\delta + \delta\partial = 1$ and $\delta^2 = 0$.

Let $C^+ = \bigoplus_i C_{2i}$ and $C^- = \bigoplus_i C_{2i+1}$. Then $\partial + \delta: C^+ \to C^-$ is an isomorphism of free Λ-modules, since $(\partial+\delta)^2 = \partial^2 + \partial\delta + \delta\partial + \delta^2 = 1$. Using the bases of C^+, C^-, $\partial+\delta$ determines an element $\tau(C)$ in $\overline{K}_1(\Lambda)$, called the <u>torsion of C</u>. The torsion is independent of δ. We can in fact define $\tau(C)$ if $H_i(C)$ is free, as in [D20].

Let $0 \to C' \to C \to C'' \to 0$ be an exact sequence of chain complexes as above, and let H be the long exact sequence of the above sequence, regarded as a chain complex. Then H is acyclic.

<u>Lemma 1</u>. <u>$\tau(C) = \tau(C') + \tau(C'') + \tau(H)$</u>.

Proof: See Milnor [D20], Theorem 3.2.

With minor modifications, we can do all of the above s-free Λ-modules.

<u>Definition</u>. A chain complex C is <u>simple</u> if $\tau(C) = 0$. If $F: C \to C'$ is a chain homotopy equivalence, then its mapping cone $\overline{C}$ is free and acyclic. We say F is a <u>simple equivalence</u> if $\tau(\overline{C}) = 0$.

When Λ has an involution $*$, then there is an induced involution on $K_1(\Lambda)$ also denoted $*$, defined by

$$(a_{ij})^* = (a_{ji}{}^*).$$

Define the <u>Steinberg group</u> $St(n,\Lambda)$ for $n \geq 3$ to be the group with generators x_{ij}^{λ}, $1 \leq i \neq j \leq n$, $\lambda \in \Lambda$, and with relations

$$x_{ij}^{\lambda}\, x_{ij}^{\mu} = x_{ij}^{\lambda+\mu}$$

$$[x_{ij}^{\lambda}, x_{j\ell}^{\mu}] = x_{i\ell}^{\lambda\mu},\ i \neq \ell$$

$$[x_{ij}^{\lambda}, x_{k\ell}^{\mu}] = 1,\ j \neq k,\ i \neq \ell.$$

There are natural inclusions $St(n,\Lambda) \subset St(n+1,\Lambda)$; let $St(\Lambda) = \bigcup_n St(n,\Lambda)$. Define $\phi: St(\Lambda) \to E(\Lambda)$ by $\phi(x_{ij}^{\lambda}) = e_{ij}^{\lambda}$, and let $K_2(\Lambda) = \ker(\phi: St(\Lambda) \to E(\Lambda))$.

<u>Lemma 2</u>. $K_2(\Lambda) \cong H_2(E(\Lambda);\mathbb{Z})$.

Proof: Milnor [C8], Theorem 5.10.

1.8. <u>Localization</u>.

Let Λ be a ring and $C(\Lambda)$ its center, $C(\Lambda) = \{\lambda \in \Lambda \mid \lambda\mu = \mu\lambda$ for every $\mu \in \Lambda\}$. Let $S \subseteq C(\Lambda) - 0$ be multiplicatively closed and contain 1.

Define the <u>localization of Λ away from S</u>, $S^{-1}\Lambda$, to be $\Lambda \times S$ modulo the relation $(\lambda,s) \sim (\mu,t)$ if there exists $u \in S$ so that $(\lambda t - \mu s)u = 0$. Let λ/s denote the equivalence class of (λ,s). Define a ring structure on $S^{-1}\Lambda$ by $\lambda/s + \mu/t = \lambda t+\mu s/st$, $\lambda/s \cdot \mu/t = \lambda\mu/st$. There is a natural ring homomorphism $\Lambda \to S^{-1}\Lambda$ given by $\lambda \mapsto \lambda/1$.

If M is a Λ-module, we define $S^{-1}M = M \otimes_{\Lambda} S^{-1}\Lambda$, where $S^{-1}\Lambda$ has a left Λ-module structure defined by the map $\Lambda \to S^{-1}\Lambda$.

Lemma 1. If $M' \to M \to M''$ is exact, then $S^{-1}M' \to S^{-1}M \to S^{-1}M''$ is also exact.

Proof: The proof given in Atiyah-MacDonald [A1], Proposition 3.3, works since $S \subseteq C(\Lambda) - 0$.

Examples: (1) Let $I \subset \Lambda$ be a prime ideal. Then $\Lambda - I$ is multiplicatively closed and we let $\Lambda_I = (\Lambda - I)^{-1}\Lambda$, called the *localization of Λ at I.*

(2) If $A \subset \Lambda$ is a subset, let $S = \Pi(A) = \{a_1^{k_1} \ldots a_n^{k_n} \mid a_i \in A,\ k_i$ a non-negative integer$\}$. Define the *localization away from A* by $\Lambda_A = S^{-1}\Lambda$.

For example, $\mathbb{Z}_2 = \mathbb{Z}[\frac{1}{2}]$, $\mathbb{Z}_{(2)} = \mathbb{Z}[\frac{1}{3}, \frac{1}{5} \ldots]$, $\mathbb{Q} = \mathbb{Z}_P$ where P is the set of all primes.

Definition. Let X be a CW-complex and P a set of primes. A *localization* of X is a space X_P and a map $f: X \to X_P$ so that f localizes homology, i.e.

$$\begin{array}{ccc} \tilde{H}_*(X) & \xrightarrow{f_*} & \tilde{H}_*(X_P) \\ & \searrow \quad \swarrow \cong & \\ & \tilde{H}_*(X) \otimes \mathbb{Z}_P & \end{array}$$

commutes.

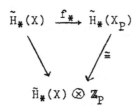

Equivalently, f localizes homotopy. (Sullivan [G15], Theorem 2.1).

A space Y is said to have <u>local homology</u> if $\tilde{H}_*(Y)$ is a $\mathbb{Z}_P$-module.

Let $k_1, k_2, \ldots$ be an enumeration of $\Pi(P)$ and choose a map $d_i : S^n \to S^n$ of degree k_i. Define the <u>local n-sphere</u> S_P^n to be the infinite telescope $M_{d_1} \cup M_{d_2} \cup \ldots$ of mapping cylinders. The inclusion $S^n \to M_{d_1} \to S_P^n$ localizes homology.

<u>Theorem 1. (Sullivan [G15])</u>. <u>Let X be a CW-complex with one 0-cell and no 1-cells</u> (e.g. $\pi_1(X) = 0$). <u>Then X has a localization</u>.

Proof: If X is a 2-complex, then $X \simeq \vee S^2$, a wedge of 2-spheres, and $X \to X_P = \vee S_P^2$ is a localization.

Inductively assume the theorem is true for complexes of dimension $\leq n-1$. Let $X^{(i)}$ be the i-skeleton; write $X^{(n)} = X^{(n-1)} \cup_f c(\vee S^{n-1})$, where c denotes cone and $f : \vee S^{n-1} \to X^{(n-1)}$ is the attaching map. We can extend f to $f_P : \vee S_P^{n-1} \to X_P^{(n-1)}$, and let $X_P^{(n)} = X_P^{(n-1)} \cup_{f_P} c(\vee S_P^{n-1})$.

Then we have by the Puppe sequence ([A25]), an exact ladder

$$\begin{array}{ccccccccc}
VS^{n-1} & \xrightarrow{f} & X^{(n-1)} & \rightarrow & X^{(n)} & \rightarrow & VS^{n} & \xrightarrow{Sf} & SX^{(n-1)} \\
\downarrow & & \downarrow & & \downarrow & & \downarrow & & \downarrow \\
VS_{\underline{p}}^{n-1} & \xrightarrow[f_{\underline{p}}]{} & X_{\underline{p}}^{(n-1)} & \rightarrow & X_{\underline{p}}^{(n)} & \rightarrow & VS_{\underline{p}}^{n} & \xrightarrow[Sf_{\underline{p}}]{} & SX_{\underline{p}}^{(n-1)}
\end{array} .$$

Since all spaces on the bottom row except $X_{\underline{p}}{}^{(n)}$ have local homology, $X_{\underline{p}}{}^{(n)}$ must have local homology, and it follows that $X^{(n)} \rightarrow X_{\underline{p}}{}^{(n)}$ is a localization. Now define $X_{\underline{p}} = \bigcup_{n} X_{\underline{p}}{}^{(n)}$, which gives the result.

The following theorem is useful in studying torsion; it follows in the same manner as Theorem 1, but by considering only those cells in Y not in X.

Theorem 2. Let (Y,X) be a CW-pair with inclusion inducing $\pi_1(X) \cong \pi_1(Y)$. Then there is a pair $((Y,X)_{\underline{p}},X)$ and a map $f:(Y,X) \rightarrow ((Y,X)_{\underline{p}},X)$ so that f localizes relative homology, i.e.

$$\begin{array}{ccc}
H_*(Y,X) & \xrightarrow{f_*} & H_*((Y,X)_{\underline{p}},X) \\
& \searrow \qquad \swarrow \cong & \\
& H_*(Y,X) \otimes \mathbb{Z}_{\underline{p}} . &
\end{array}$$

commutes.

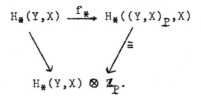

Let X be simply-connected and P a set of primes. Define the colocalization away from P, X^P, to be the fiber of the map $X \to X_{(P)}$. Similarly let $X^{(P)}$ be the fiber of $X \to X_P$. We list some elementary properties:

(a) If $\pi_i(X)$ is finite for all i, then $X^P \simeq X_{P'}$ (since then $X \simeq \prod_p X_p$).

(b) If P is the set of all primes, then $X^P \simeq *$.

(c) In general, $\pi_i(X^P) =$ (P-torsion in $\pi_i(X)) \oplus (\mathbb{Z}_{P'}/\mathbb{Z})^{rk\pi_{i+1}(X)}$, where P' is the set of primes not in P and rk denotes rank. Thus

$$\pi_i(X^P) \otimes \mathbb{Z}_P \cong \pi_i(X_P)$$

$$\pi_i(X^P) \otimes \mathbb{Z}_{(P)} = 0$$

$$\pi_i(X^P) \otimes \mathbb{Z}_q = \begin{cases} \pi_i(X) \otimes \mathbb{Z}_q & q \in P \\ (\mathbb{Z}_{P'-q}/\mathbb{Z})^{rk\pi_{i+1}(X)} & q \notin P. \end{cases}$$

(d) $(X^P)_{(P)} \simeq *$; $(X^P)_P \simeq (X_{(P)})^{(P)}$; $(X_P)^{(P)} \simeq *$.

(e) If $\pi_2(X)$ is finite, then $H_*(X,X^P) \cong \tilde{H}_*(X_{(P)})$ and $X_{(P)}$ is the cofiber of $X^P \to X$. (see [K11]).

Chapter 2. Whitehead Torsion

Let π be a multiplicative group and $w:\pi \to \{\pm 1\}$ a homomorphism; let R be a ring with 1. There is a natural ring homomorphism $j:\mathbb{Z} \to R$. Let H be the subgroup of $(R\pi)^{\bullet}$ generated by π and $\mathrm{Im}(j)\cap R^{\bullet}$, $\langle \pi, \mathrm{Im}(j)\cap R^{\bullet}\rangle$.

Definition. <u>The Whitehead group of π with coefficients in R</u> is defined by $Wh(\pi;R) = K_1(R\pi)/H$ where

$$H \subset (R\pi)^{\bullet} = GL(1;R\pi) \hookrightarrow GL(R\pi) \to K_1(R\pi).$$

More generally, if Λ is a ring and $\mathcal{F}:\mathbb{Z}\pi \to \Lambda$ is a ring homomorphism, then we can define $Wh(\mathcal{F})$. This has been done by Cappell and Shaneson [K5], though their definition differs from the one given here.

The homomorphism w induces an involution $*$ on $Wh(\pi;R)$.

Examples: (1) $Wh(1;\mathbb{Z}_p) = 0$: For $K_1(\mathbb{Z}_p) \cong \mathbb{Z}_p^{\bullet} \oplus SK_1(\mathbb{Z}_p)$

$$= \mathbb{Z}_p^{\bullet}$$

and $\mathrm{Im}(j)\cap \mathbb{Z}_p^{\bullet} = \pm\mathbb{I}(P)$, so $H = \mathbb{Z}_p^{\bullet}$.

(2) $Wh(\pi;\mathbb{Z}) = Wh(\pi)$ as defined in Milnor [D20].

(3) If π is abelian, then $Wh(\pi;R) \cong \left((R\pi)^{\bullet}/H\right) \oplus SK_1(R\pi)$.

Thus if π is an elementary abelian p-group, $p \geq 3$, of rank $n \geq 3$, then

$Wh(\pi;\mathbb{Z}_P) = (\mathbb{Z}_P\pi)^{\bullet}/\langle\mathbb{Z}_P^{\bullet},\pi\rangle$, $P \neq \phi$. If $n \leq 2$, then this is true for all P.

Definition. Let $f:X \to Y$ be a homology equivalence over R, $\pi = \pi_1(Y)$. Define the torsion of f, $\tau(f;R) \in Wh(\pi;R)$, to be the torsion of the chain complex $C_*(M_f,X) \otimes_{\mathbb{Z}\pi} R\pi$, where M_f is the mapping cylinder of f.

f is a simple homology equivalence over R if $\tau(f;R) = 0$. In case f is an inclusion $X \subset Y$, $\tau(f;R)$ is denoted $\tau(Y,X;R)$.

The following properties are easy to prove (see Milnor [D20] Lemmas 7.5 - 7.8 for the case $R=\mathbb{Z}$):

(1) $\tau(M_f,Y;R) = 0$.

(2) If $f \simeq g$ then $\tau(f;R) = \tau(g;R)$.

(3) If $g:Y \to Z$ is a homology equivalence over R, then $\tau(gf) = \tau(g) + g_*\tau(f)$, where $g_*:Wh(\pi_1(Y);R) \to Wh(\pi_1(Z);R)$ is the natural map.

As before, torsion can be defined for f if $H_*(M_f,X;R\pi)$ is free (or s-free).

The following theorem will be useful in Chapter 4.

Theorem 1. Suppose X is a finite connected CW-complex, $\pi = \pi_1(X)$, and

(1) $H_i(X;R\pi) = 0$ for $i \neq r$

(2) $H^{r+1}(X;M) = 0$ *for every* $R\pi$*-module* M.

Then $H_r(X;R\pi)$ *is a finitely generated, s-free* $R\pi$*-module. Furthermore, we may choose our bases so that* $\tau(C_*(X) \otimes R\pi;R) = 0$.

Proof: Let $C_i = C_i(X) \otimes_{\mathbb{Z}\pi} R\pi$, $Z_i = \ker(\partial_i)$, $B_i = \operatorname{Im}(\partial_{i+1})$, $H_i = H_i(C_*)$. We have exact sequences

$$0 \to Z_i \to C_i \to B_{i-1} \to 0$$
$$0 \to B_i \to Z_i \to H_i \to 0$$

which gives the exact sequence

$$0 \to Z_i \to C_i \to Z_{i-1} \to 0 \qquad i \leq r$$

since $H_i = 0$ for $i < r$.

Since $Z_0 = C_0$, Z_i is projective for $i \leq r$ by induction. Thus C_* is chain homotopic to

$$C_*' : \ldots \to C_{r+2} \to C_{r+1} \to Z_r \to 0$$

by the standard proof (Spanier [A15], Theorem 4.2.5.).

Let $i:B_r \to Z_r$ be the inclusion and $\partial'_{r+1}:C_{r+1} \to B_r$ so that $\partial_{r+1} = i\partial'_{r+1}$. This defines an element in $\operatorname{Hom}(C_{r+1},B_r)$, and if ∂ denotes the dual homomorphism ∂^*, then $\partial(\partial'_{r+1}) = \partial'_{r+1} \circ \partial_{r+2}$

$$= 0.$$

Thus ∂'_{r+1} is a cocycle. But by (1), $H_{r+1}(\mathrm{Hom}(C_*,B_r)) = 0$, so ∂'_{r+1} is a coboundary. Thus $\partial'_{r+1} = f \circ \partial_{r+1}$ for some $f:C_r \to B_r$

$$= fi\partial'_{r+1}.$$

Since ∂'_{r+1} is onto, $fi = 1$.

So B_r is a direct summand of Z_r, and $Z_r \cong B_r \oplus H_r$. Thus H_r is projective. Since $C_r \cong Z_r \oplus Z_{r-1}$ and C_r is finitely generated, H_r is finitely generated.

The complex $C_*'' : \ldots \to C_{r+2} \to C_{r+1} \to B_r \to 0$ has zero homology and so is chain contractible, since B_r is projective. Let C_*''' be $0 \to Z_r \to C_{r-1} \to \ldots \to C_0 \to 0$. Then $H(C_*''') = 0$ and so C_*''' is contractible. Let $\mathcal{H}$ be the complex $0 \to H_r \to 0$. Then

$$C_*' \simeq C_* = C_*'' \oplus C_*''' \oplus \mathcal{H} \simeq \mathcal{H}.$$

Since $C_*'' \oplus C_*'''$ is contractible, the sum of the even terms is isomorphic to the sum of the odd terms. Thus

$$B_r \oplus Z_{r-1} \oplus \bigoplus_{i \neq 0} C_{r+2i} \cong \bigoplus_i C_{r+2i+1}.$$

Adding H_r to both sides,

$$\bigoplus C_{r+2i} \cong H_r \oplus \bigoplus C_{r+2i+1}.$$

Thus H_r is s-free.

We may choose an s-basis for H_r so that $\tau(C_*;R) = 0$ by Lemma 1.7.1.

The result remains true for a finite CW-pair (Y,X).

We now give a geometric characterization of torsion over $\mathbb{Z}_P$. Let $r \in \Pi(P)$ and fix a map $\phi_r^n : S^{n-1} \to S^{n-1}$ of degree r; let $C^n(r)$ denote the mapping cone of ϕ_r^n, $C_{\phi_r^n}$.

Definition: Let (Y,X) be a finite CW-pair and suppose

$$Y = X \cup C^n(r) \cup C^{n+1}(r)$$

so that there exists a map $f : C_{\hat{\phi}_r^n} \to Y$, where

$\hat{\phi}_r^n : (D^n, S^{n-1}) \to (D^n, S^{n-1})$ is of degree r, satisfying:

(1) $f | D^n \cup C_{\hat{\phi}_r^n | S^{n-1}}$ is the attaching map for $C^{n+1}(r)$,

(2) $f | S^{n-1}$ is the attaching map for $C^n(r)$, and

(3) $f(D^n) \subset X$.

We say X is obtained from Y by <u>elementary P-collapse</u>, $Y \overset{P}{\searrow} X$, or Y is obtained from X by <u>elementary P-expansion</u>, $X \overset{P}{\nearrow} Y$.

We write $X \underset{P}{\wedge\!\!\searrow} Y$ if there exists a sequence $X = Y_0, Y_1, \ldots, Y_m = Y$ so that either $Y_i \overset{P}{\searrow} Y_{i+1}$ or $Y_i \overset{P}{\nearrow} Y_{i+1}$, and we say there is a <u>formal deformation over $\mathbb{Z}_P$</u> from X to Y.

<u>Lemma 1.</u> <u>If</u> $Y \overset{P}{\searrow} X$, <u>then</u> $\pi_i(Y,X) = 0$ <u>for</u> $i \neq n$ <u>and</u> $\pi_n(Y,X) \otimes \mathbb{Z}_P = 0$.

A <u>deformation over</u> $\mathbb{Z}_p$ is a sequence of maps $\{f_0,\ldots,f_m\}$, $f_i:Y_i \to Y_{i+1}$ the inclusion if $Y_i \overset{P}{\nearrow} Y_{i+1}$, $f_i:Y_{i+1} \to Y_i$ the inclusion if $Y_i \overset{P}{\searrow} Y_{i+1}$, $Y_0 = X$, $Y_m = Y$. If Z is a subcomplex of each Y_i and $f_i|Z = 1$, then we say f is a deformation over $\mathbb{Z}_p$ <u>relative</u> to Z.

<u>Lemma 2.</u> <u>Let</u> (Y,X) <u>be a connected CW-pair so that the inclusion</u> $X \subset Y$ <u>is a homology equivalence over</u> $\mathbb{Z}_p$. <u>Then</u> $Y \underset{P}{\curvearrowright} Z$ rel X <u>where</u> $Z = X \cup \bigcup_{i=1}^{k} C^n(r_i) \cup \bigcup_{i=1}^{k} C^{n+1}(s_i)$.

Proof: Inductively assume $Y \underset{P}{\curvearrowright} Z_m$ rel X where $Z_m = X \cup C^m(r) \cup$ cells of type $C^t(s)$, $t \geq m$. Since $\pi_m(Y,X) \otimes \mathbb{Z}_p = 0$, $\pi_m(Z_m,X) \otimes \mathbb{Z}_p = 0$. It follows that there exists a map $f:C^{m+1}(r) \to Z_m$ extending the map $C^m(r) \to Z_m$ so that $f(\partial_- D^{m+1}) \subset X$. Let $W = Z_m \cup_f C^{m+2}(r)$ where $C^{m+1}(r) \subset C^{m+2}(r)$ is the natural inclusion. Clearly $Z_m \overset{P}{\nearrow} W$. Here we let $D_-{}^k$ (resp. $D_+{}^k$) denote the lower (resp. upper) part of the standard k-disc, and $\partial_- D^k = \partial D_-{}^k$, $\partial_+ D^k = \partial D_+{}^k$.

Let $E^{m+1} = \partial_+ D^{m+2} \subset C^{m+2}(r) \subset W$, and $W' = X \cup C^m(r) \cup E^{m+1}$. Then $W' \overset{P}{\searrow} X$; so $Y \underset{P}{\curvearrowright} Z_m \underset{P}{\nearrow} W \underset{P}{\curvearrowright} X \cup C^{m+2}(r) \cup$ higher cells, since $W' \overset{P}{\searrow} X$. Thus by induction,

$$Y \underset{p}{\curvearrowright} Z = X \cup \bigcup_{i=1}^{k} C^n(r_i) \cup \bigcup_{i=1}^{\ell} C^{n+1}(s_i).$$

Since $H_*(Y,X;\mathbb{Z}_p\pi) = 0$, we must have $k=\ell$.

<u>Lemma 3</u>. <u>Suppose</u> $Y = X \cup C^n(r)$, $n \geq 2$, $\pi = \pi_1(Y)$. <u>Then</u> $\pi_n(Y,X) \otimes \mathbb{Z}_p$ <u>is a free</u> $\mathbb{Z}_p\pi$-<u>module with basis</u> $[\hat{\phi}_r^{\,n}] \otimes 1$, <u>where</u> $\hat{\phi}_r^{\,n}: (D^n, S^{n-1}) \to (C^n(r), S^{n-1}) \downarrow (Y,X)$ <u>extends</u> $\phi_r^{\,n}$.

Proof: It follows easily that $X \subset Y$ induces an isomorphism of fundamental groups. We have that $H_n(Y,X;\mathbb{Z}_p\pi) = H_n(\tilde{Y},\tilde{X};\mathbb{Z}_p)$ is a free $\mathbb{Z}_p\pi$-module with basis $(\tilde{\phi}_r^{\,n})_*(\iota_n)$ where $\iota_n \in H_n(D^n,S^{n-1};\mathbb{Z}_p)$ is a generator and $\tilde{\phi}_r^{\,n}: (D^n,S^{n-1}) \to (\tilde{Y},\tilde{X})$ is a lift of $\hat{\phi}_r^{\,n}$.

The Hurewicz map $\pi_n(\tilde{Y},\tilde{X}) \to H_n(\tilde{Y},\tilde{X})$, $[\psi] \mapsto \psi_*(\iota_n)$, is an isomorphism of $\mathbb{Z}\pi$-modules, and so

$$H_n(\tilde{Y},\tilde{X};\mathbb{Z}_p) \cong H_n(\tilde{Y},\tilde{X}) \otimes \mathbb{Z}_p \cong \pi_n(\tilde{Y},\tilde{X}) \otimes \mathbb{Z}_p \cong \pi_n(Y,X) \otimes \mathbb{Z}_p$$

The lemma is true for $n=2$ provided the attaching map $S^{n-1} \to X$ is a point.

Suppose (Y,X) is a finite CW-pair with $X \subset Y$ a homology equivalence over $\mathbb{Z}_p$. Then by Lemma 2, we can assume $Y = X \cup \bigcup_{i=1}^{k} C^n(r_i) \cup \bigcup_{i=1}^{k} C^{n+1}(s_i)$. Let $Y_1 = X \cup \bigcup_{i=1}^{k} C^n(r_i)$. By Lemma 3, $\pi_n(Y_1,X) \otimes \mathbb{Z}_p$ and

$\pi_{n+1}(Y,Y_1) \otimes \mathbb{Z}_P$ are free $\mathbb{Z}_P\pi$-modules with generators $[\phi_1],\ldots,[\phi_k]$, and $[\psi_1],\ldots,[\psi_k]$ respectively, where ϕ_i,ψ_i are the attaching maps.

Definition. The <u>matrix of</u> (Y,X) (with respect to the maps ϕ_i,ψ_i) is defined to be the matrix of the map $\pi_{n+1}(Y,Y_1) \otimes \mathbb{Z}_P \to \pi_n(Y_1,X) \otimes \mathbb{Z}_P$ coming from the exact sequence of the triple (Y,Y_1,X). This matrix is invertible since $\pi_*(Y,X) \otimes \mathbb{Z}_P = 0$.

<u>Lemma 4. Suppose the matrix of</u> (Y,X) <u>is the identity for some choice of</u> ϕ_i,ψ_i. <u>Then</u> $Y \underset{P}{\nearrow\!\!\searrow} X$ rel X.

Consider the following "elementary" operations on an invertible matrix A over $\mathbb{Z}_P\pi$, where R_i,C_i denote the i-th row, column of A:

(1) $R_i \mapsto \pm agR_i \qquad g \in \pi,\ a \in \Pi(P)$

(2) $C_i \mapsto \pm aC_ig$

(3) $R_i \mapsto R_i + xR_j \qquad x \in \mathbb{Z}_P\pi$

(4) $C_i \mapsto C_i + C_jx$

(5) $A \mapsto \begin{pmatrix} A & 0 \\ 0 & I \end{pmatrix}$.

If A can be transformed to B by operations of type (1) - (4), then $A = CBD$ where C,D are elementary matrices.

Lemma 5. *Let* A *be the matrix of* (Y,X) *and suppose* A *can be transformed to the identity by operations of type* (1) - (5). *Then* $Y \underset{P}{\nearrow\!\searrow} X$.

Proof: It follows by elementary homotopy constructions that if B is obtained from A by operations of type (1), (3) or (5), then there exists Z so that $Y \underset{P}{\nearrow\!\searrow} Z$ rel X and the matrix of (Z,X) is B.

Suppose A can be transformed to I by operations of type (1) - (5). Then

$$I = C \begin{pmatrix} A & 0 \\ 0 & I \end{pmatrix} D, \quad C, D \text{ elementary, and so}$$

$$I = DC \begin{pmatrix} A & 0 \\ 0 & I \end{pmatrix}.$$ Therefore there exists Z so that $Y \underset{P}{\nearrow\!\searrow} Z$ rel X and the matrix of (Z,X) is I. By Lemma 4, $Z \underset{P}{\nearrow\!\searrow} X$ rel X, and so $Y \underset{P}{\nearrow\!\searrow} X$ rel X.

Definition. Let X and Y be finite CW-complexes. Then X and Y have the same *simple homology type over* $\mathbb{Z}_P$ if there is a sequence $X = Z_0, Z_1, \ldots, Z_r = Y$, each Z_i a finite complex and simple homology equivalences over $\mathbb{Z}_P$, $Z_i \to Z_{i+1}$ or $Z_{i+1} \to Z_i$.

This gives the geometric characterization of torsion:

Theorem 2. Suppose (Y,X) *is a finite CW pair with* $X \subset Y$ *a homology equivalence over* $\mathbb{Z}_p$. *Then* $\tau(Y,X;\mathbb{Z}_p) = 0$ *iff* $Y \underset{p}{\nearrow\!\!\searrow} X$ rel X.

Corollary 1. If X *and* Y *are finite CW-complexes, then* X *and* Y *have the same simple homology type over* $\mathbb{Z}_p$ *if and only if there is a deformation from* X *to* Y.

Definition. Let $(N;M_+,M_-)$ be a manifold triad, $\partial N = M_+ \cup M_-$, M_+,M_- closed. Then N is an *h-cobordism over* $\mathbb{Z}_p$ between M_+ and M_- provided the inclusions $M_+ \subset N$, $M_- \subset N$ are homology equivalences over $\mathbb{Z}_p$.
N is called an *s-cobordism over* $\mathbb{Z}_p$ if $\tau(N,M_+;\mathbb{Z}_p) = 0$.

Theorem 3. Let M^n be a closed manifold, $n \geq 5$, and $x \in Wh(\pi_1 M; \mathbb{Z}_p)$. Then there exists an h-cobordism over $\mathbb{Z}_p$ $(N;M,M')$ with $\tau(N,M;\mathbb{Z}_p) = x$.

Proof: Let x be represented by a $k \times k$ matrix A, and let N_1 be $M \times I \cup k$ 2-handles, where the attaching maps $S^1 \times D^{n-1} \to M \times 1$ are trivial. Let $M_1 = \partial_+ N_1$ $(M = \partial_- N_1)$.

Each row of A, R_i, represents an element of $\pi_2(N_1, M)$. Since $N_1 = M_1 \times I \cup (n-2)$-handles, and $n-2 > 2$, R_i represents an element in $\pi_2(M_1)$. By general position and Milnor [D20], Theorem 11.1, R_i is represented by a trivial embedding. Use these embeddings to attach 3-handles to N_1 to get N. This is the desired h-cobordism over $\mathbb{Z}_p$. See also Lemma 6.2.1.

Chapter 3. Poincare Complexes

3.1. Poincare Duality

Let X be a finite CW-complex with basepoint $*$, $\pi = \pi_1(X,*)$, and $w:\pi \to \{\underline{+}1\}$ a homomorphism. Given $c \in C_n(X) \otimes_\Lambda \mathbb{Z}$, $\Lambda = \mathbb{Z}\pi$, and π acts trivially on $\mathbb{Z}$, define $c \cap : C^q(X) \to C_{n-q}(X)$ to be cap product with $tr(c)$, where $tr: C_n(X) \otimes \mathbb{Z} \to C_n(X)$ is the transfer ([A2], pg. 243). If π is infinite, $tr(c)$ is an infinite chain, but since we are using compact supports, the same formula holds.

Definition X is a Poincare complex over R if there exists a fundamental class $[X] \in H_n(X;\mathbb{Z})$ so that $[X] \cap : H^q(X;R\pi) \to H_{n-q}(X;R\pi)$ is an isomorphism. The dimension of X is n.

If X is a Poincare complex over R, then since $C_*(X) \otimes_\Lambda R\pi$ and $\mathrm{Hom}_\Lambda(C_*(X), R\pi)$ are chain complexes of free $R\pi$-modules, $c \cap : \mathrm{Hom}_\Lambda(C_*(X), R\pi) \to C_*(X) \otimes_\Lambda R\pi$ is a chain equivalence, where c is a representative cycle for $[X]$. The free module $C_q(X)$ has a basis given by choosing lifts of cells in X; we give $C_q^*(X)$ the dual basis.

Definition The torsion of X over R is defined to be the image of $\tau(D_*;R) \in \overline{K}_1(R\pi)$ in $Wh(\pi;R)$, where D_* is the mapping cone of $c \cap$. X is a simple Poincare complex over R if its torsion over R is zero.

Theorem 1. If M^n is a closed manifold, then M is a simple Poincare complex of dimension n over any ring.

Proof: Our proof gives a more general result. Let Y be a finite CW-complex and $\mathcal{S}$ a sheaf over Y. Define the sheaf of local homology groups $\mathcal{H}_*(Y;\mathcal{S})$ by the presheaf $U \mapsto H_*(U;\mathcal{S})$. The stalks of $\mathcal{H}_*(Y;\mathcal{S})$ are given by $\mathcal{H}_*(Y;\mathcal{S})_y = H_*(Y,Y-y;\mathcal{S})$. According to Bredon [A2], pg. 208, there exists a spectral sequence

$$E_2^{p,q} = H^p(X;\mathcal{H}_q(Y;\mathcal{S})) \Rightarrow H_{q-p}(Y;\mathcal{S}) .$$

Y is called a homology manifold over R of dimension n if $\mathcal{H}_p(Y;R) = 0$ for $p \neq n$ and $\mathcal{O} = \mathcal{H}_n(Y;R)$ is locally constant with stalks isomorphic to R. In particular, M is a homology manifold over R.

Let $B = R\pi$, $\pi = \pi_1(Y)$, and $\mathcal{B}$ as in Section 1.2. If Y is a homology manifold over R, then

$$H^p(Y;\mathcal{B}^t) \cong H^p(Y;\mathcal{B}\otimes\mathcal{O})$$

$$\cong E_2^{p,n} \cong E_\infty^{p,n} \cong H_{n-p}(Y;\mathcal{B}) .$$

This isomorphism is in fact given by cap product and is simple on the chain level. See Bredon [A2], Corollary 10.2,

Wall [H19], Theorem 2.1.

Let $A \subset Wh(\pi;R)$ be a <u>conjugate-closed</u> subgroup, i.e., $A^* = A$. (A is also called <u>self-dual</u>.)

<u>Corollary 1. If M^n is a manifold and $f:M \to X$ is a homology equivalence over R so that $\tau(f;R) \in A$, then X is a Poincare complex over R with torsion lying in A</u>.

Definition. A finite CW-pair (Y,X) is a <u>Poincare pair over R</u> of dimension n if there is $[Y,X] \in H_n(Y,X;Z)$ so that

$$[Y,X] \cap : H^q(Y;R\pi) \to H_{n-q}(Y,X;R\pi)$$

is an isomorphism, $\pi = \pi_1(Y)$, and X is a Poincare complex over R so that $\partial[Y,X] = [X]$. A map $f:X \to Y$ between Poincare complexes is of <u>degree 1</u> if $f_*[X] = [Y]$.

<u>Theorem 2. Let X,Y be Poincare complexes over R of dimension n and $f:X \to Y$ a 1-connected degree 1 map. Then</u>

(a) $[X] \cap : K^q(X;R\pi) \to K_{n-q}(X;R\pi)$ <u>is an isomorphism, and</u>

(b) <u>there exist split short exact sequences</u>

$$0 \to K_q(X;R\pi) \to H_q(X;R\pi) \xrightarrow{f_*} H_q(Y;R\pi) \to 0$$

$$0 \to H^q(Y;R\pi) \xrightarrow{f^*} H^q(X;R\pi) \to K^q(X;R\pi) \to 0 .$$

Proof: (a) follows from the naturality of cap product. For (b), the composition

$$H_q(X;R\pi) \xrightarrow{f_*} H_q(Y;R\pi) \xrightarrow{([Y]\cap)^{-1}} H^{n-q}(Y;R\pi) \xrightarrow{f^*} H^{n-q}(X;R\pi) \xrightarrow{[X]\cap} H_q(X;R\pi)$$

is the identity. Thus f_* is surjective and f^* is injective; this also defines the splittings.

A similar result holds in the relative case. Now suppose X and Y have torsion in $A \subset Wh(\pi;R)$. Define $C_*(f) = C_*(M_f,X)$, and assume (M_f,X) satisfies the hypothesis of Theorem 2.1.

Remark. Let $q = n - r + 1$. We have $H^q(f;R\pi) \cong K^q(X;R\pi)$ and $H_r(f;R\pi) = K_r(X;R\pi)$ and furthermore,

$$[X]\cap : K^q(N;R\pi) \to K_{r-1}(N;R\pi)$$

is an isomorphism with torsion in A (on the chain level).

This also is true in the relative case, with $C_*(f) = C_*(M_f, X \cup \partial Y)$.

Theorem 3. Let X be a Poincare complex over $\mathbb{Z}_p$ of dimension $n \geq 5$. Then there is a Poincare complex over $\mathbb{Z}_p$, X' , and a simple homology equivalence over $\mathbb{Z}_p$ $X \to X'$, so that $X' = X_o \cup_{\partial M} M$ where M is a manifold obtained from D^n by adding 1-handles, and $\dim X_o \leq n - 2$. Furthermore, $\pi_1(M) \to \pi_1(X')$ is surjective.

Proof: Assume X is connected. Then we can assume X has one 0-cell ℓ^o, and say k 1-cells $\ell_i^{'}$ with both ends attached at ℓ^o. We have $\partial\ell_i^{'} = \alpha_i\ell^o - \ell^o$, where $\alpha_i \in \pi_1(X)$ is defined by $\ell_i^{'}$. By duality, we can assume there exists one n-cell ℓ^n and k $(n - 1)$-cells ℓ_i^{n-1}, so that $\partial\ell^n = \sum_{i=1}^{k} w(\alpha_i)\ell_i^{n-1} - \ell_i^{n-1}$ ($\mathbb{Z}_p\pi$ coefficients).

Define X' as $X_o \bigcup_{\partial M} M$, where X_o is the $(n - 2)$-skeleton of X, $M = D^n \cup k$ 1-handles, so that $\partial\ell^n = \Sigma w(\alpha_i)\alpha_i\ \ell_i^{n-1} - \ell_i^{n-1}$ with $\mathbb{Z}\pi$-coefficients.

If $(X,\partial X)$ is a Poincare pair over $\mathbb{Z}_p$, then the theorem remains true, with $(X,\partial X)$ having the $\mathbb{Z}_p$-homology type of $(X',\partial X)$, where $X' = X_o \bigcup_{\partial M} M$, $\partial X \subset X_o$, $\dim(X_0 - \partial X) \leq n - 2$.

Theorem 4. Let $(Y;X_+,X_-)$ be a Poincare triad over R of dimension n so that the inclusions $X_+ \subset Y$, $X_- \subset Y$ are homology equivalences over R. Assume $C_*(Y,X_+) \otimes R$ is simply chain equivalent to a complex with 2 non-zero terms. Then $\tau(Y,X_+;R) = (-1)^{n-1}\ \tau(Y,X_-;R)^*$.

Proof: Let M be an $R\pi$-module and $f:M \to M$ a homomorphism with matrix $A = (a_{ij})$. Then the dual homomorphism $f^*:M^* \to M^*$ given by $f^*(x)(m) = x(f(m))^*$ has matrix $A^{*t} = (a^*_{ji})$.
We may assume $C_*(Y,X_+) \otimes R$ has the form
$0 \to C_k \overset{\partial}{\to} C_{k-1} \to 0$. Since $X_+ \subset Y$ is a homology equivalence over R, ∂ is invertible. Furthermore, the matrix of ∂ represents $(-1)^{k-1}\tau(Y,X_+;R)$ in $Wh(\pi;R)$.

Dually, $0 \to C^*_{n-k+1} \overset{\partial^*}{\to} C^*_{n-k} \to 0$ is chain equivalent to $C_*(Y,X_-) \otimes R\pi$, and so the matrix of ∂^* represents $(-1)^{n-k}\tau(Y,X_-;R)$ in $Wh(\pi;R)$. Since the involution $*$ on $Wh(\pi;R)$ sends A to A^{*t},
$(-1)^{k-1}\tau(Y,X_+;R) = (-1)^{n-k}\tau(Y,X_-;R)^*$.

3.2. Spherical Fibrations and Normal Maps

If ξ is a R-spherical fibration (i.e. the fiber is homology equivalent over R to a sphere) $p:E \to X$, then define the Thom space of ξ by $T(\xi) = C(E) \underset{p}{\cup} X$, where $C(E)$ is the cone of E.

We now prove a generalization of the theorem of Spivak [E19].

Theorem 1. Let $(X, \partial X)$ be a Poincare pair over $\mathbb{Z}_p$ of dimension n. Then there exists an $\mathbb{Z}_p$-spherical fibration ξ over X, with fiber homology equivalent to S^{k-1} over $\mathbb{Z}_p$, so that the generator of $H_{n+k}(T(\xi), T(\xi|\partial X); \mathbb{Z}_p)$ is spherical. The fibration ξ is unique up to stable fiber homology equivalence over $\mathbb{Z}_p$.

Proof: Let N be a regular neighborhood of X in D^{n+k}, $k \geq 4$, so that $M = N \cap S^{n+k-1}$ is a regular neighborhood of ∂X in S^{n+k-1}. Let $N_o = \overline{\partial N - M}$; then $N_o \cap M = \partial N_o = \partial M$. We have N is an $(n+k)$-manifold and $\pi_i(N - X) \to \pi_i(N)$ is an isomorphism for $i < k - 1$ and onto for $i = k - 1$. Since $N_o \to N - X$ is a homotopy equivalence, $\pi_i(N, N_o) = 0$ for $i \leq k - 1$.

Let $t':E' \to N$ be a fibration equivalent to the inclusion $N_o \to N$. If F is the fiber of t', then $\pi_1(F) \cong \pi_2(N, N_o) = 0$.

Let $\pi = \pi_1(X)$. Then there exists $[N,\partial N] \in H_{n+k}(N,\partial N;\mathbb{Z})$ so that $[N,\partial N] \cap : H^p(N,N_o) \to H_{n+k-p}(N,M)$ is an isomorphism. Since $(N,M) \simeq (X,\partial X)$, there exists $[X,\partial X] \in H_n(N,M;\mathbb{Z})$ so that $[X,\partial X] \cap : H^p(N,M;\mathbb{Z}_P\pi) \to H_{n-p}(N;\mathbb{Z}_P\pi)$ is an isomorphism. Choose $U \in H^k(N,N_o;\mathbb{Z})$ so that $[N,\partial N] \cap U = [X,\partial X]$. Then $\cup U: H^p(N;\mathbb{Z}_P\pi) \to H^{p+k}(N,N_o;\mathbb{Z}_P\pi)$ is an isomorphism, since for $y \in H^p(N;\mathbb{Z}_P\pi)$, $[N,\partial N] \cap (y \cup Y) = ([N,\partial N] \cap U) \cap y = [X,\partial X] \cap y$. Now consider the fibration of universal covers $\tilde{t}':\tilde{E}' \to \tilde{N}$ with fiber F . Then there exists $\tilde{U} \in H^k(\tilde{t}')$ so that $\cup\,\tilde{U}: H^p(\tilde{N};\mathbb{Z}_P) \cong H^{p+k}(\tilde{t}';\mathbb{Z}_P)$. It follows by a spectral sequence argument that $F \simeq S^{k-1}$ mod C , where C is the Serre class of abelian groups with exponent $p_1^{n_1}\ldots p_r^{n_r}$, $p_i \in P$. (A proof is given in Browder [G5], Lemma I.4.3. for $P = \emptyset$). Thus F is homology equivalent over $\mathbb{Z}_P$ to S^{k-1} . Let $f:E \to X$ be the pullback of t' by the inclusion $X \subset N$. This is the Spivak fibration; denote it by ξ .

We have $T(\xi) = N/N_o$ and $T(\xi|\partial X) = M/\partial M$, and the collapse $(D^{n+k},S^{n+k-1}) \to (N/N_o, M/\partial M)$ defines $\alpha \in \pi_{n+k}(T(\xi),T(\xi|\partial X))$. The the Hurewicz map

$$\pi_{n+k}(T(\xi),T(\xi|\partial X)) \otimes \mathbb{Z}_P \to H_{n+k}(T(\xi),T(\xi|\partial X);\mathbb{Z}_P)$$

sends $\alpha \otimes 1$ to a generator.

To show uniqueness, suppose $\xi_1 : E_1 \to X$, $\xi_2 : E_2 \to X$ both satisfy the theorem (assume $\partial X = \emptyset$ for simplicity). Stably, form the bundle $\xi_2 \times (\xi_1 - \xi_2)$ over $X \times X$ and consider the map F defined by

$$S^{n+k} \xrightarrow{\alpha_1} T(\xi_1) \xrightarrow{\Delta} T(\xi_2 \times (\xi_1 - \xi_2)) = T(\xi_2) \wedge T(\xi_1 - \xi_2) .$$

The map Δ is defined by the diagonal $X \to X \times X$. As in Wall [E21], Theorem 3.3, F is a duality map over $\mathbb{Z}_p$ and so $T(\xi_1 - \xi_2)$ is co-reducible over $\mathbb{Z}_p$, i.e., there is a map $T(\xi_1 - \xi_2) \to S^m$ which induces isomorphisms on cohomology with $\mathbb{Z}_p$-coefficients in degrees $\leq m$. Let $E \to X$ be the fibration obtained by suspending each fiber in $\xi_1 - \xi_2$. Then $T(\xi_1 - \xi_2)$ is obtained by identifying suspension points in E and so E is fiber homologically trivial over $\mathbb{Z}_p$. Therefore ξ_1 and ξ_2 have the same fiber homology type over $\mathbb{Z}_p$.

In Section 1.3, we constructed monoids $G_n(\mathbb{Z}_p)$ and classifying spaces $BG_n(\mathbb{Z}_p)$ for principal $G_n(\mathbb{Z}_p)$-bundles, equivalently for fibrations with fiber homology equivalent to S^{n-1} over $\mathbb{Z}_p$. Such a fibration is equivalent to a fibration with fiber S_p^{n-1}, and according to Sullivan [G15] Corollary 4.3., these fibrations are classified by the localized space $(BG_n)_p$, $G_n = G_n(\mathbb{Z})$. Thus $BG_n(\mathbb{Z}_p) \simeq (BG_n)_p$.

Let SG_n be the submonoid of G_n defined by homotopy equivalences of degree 1. Then BSG_n is the universal cover of BG_n and stably, $BG \simeq RP^\infty \times BSG$. Since the homotopy groups of BSG are finite $(\pi_i(BSG) \cong \lim\limits_{\overrightarrow{k}} \pi_{i+k+1}(S^k)$, $i > 1)$,

$$(BG)_P \simeq K(\mathbb{Z}_P^{\cdot},1) \times \prod_{p \notin P} (BSG)_{(p)} .$$

Let H = TOP, PL or O. Then there is a natural map $BH \to (BG)_P$. Let G_P/H denote the fiber of this map.

<u>Theorem 2. Let</u> SG_P/SH <u>be the fiber of</u> $BSH \to (BSH)_P$. <u>Then</u> $G_P/H \cong (SG_P/SH) \times \mathbb{Z}_P^{\cdot +}$ <u>and</u>

$$\pi_i(SG_P/SH) \otimes \mathbb{Z}_P \cong \pi_i(G/H) \otimes \mathbb{Z}_P,$$

$$\pi_i(SG_P/SH) \otimes \mathbb{Z}_{(P)} \cong \pi_i(BSH) \otimes \mathbb{Z}_{(P)} .$$

Proof: We have an exact ladder

$$\begin{array}{ccccccc}
\cdots \to & \pi_i(SG_P/SH) & \to & \pi_i(BSH) & \to & \pi_i((BSG)_P) & \to \cdots \\
 & \downarrow & & \downarrow & & \downarrow & \\
\cdots \to & \pi_i(G_P/H) & \to & \pi_i(BH) & \to & \pi_i((BG)_P) & \to \cdots
\end{array}$$

and it follows that $\pi_i(SG_P/SH) \cong \pi_i(G_P/H)$ for $i > 1$. Furthermore, $\pi_1(SG_P/SH) = \pi_1(G_P/H) = 0$, and $\pi_0(SG_P/SH) = 0$. Thus the sequence reduces to

$$0 \to \pi_1(BH) = \mathbb{Z}/2\mathbb{Z} \to \pi_1((BG)_P) = \mathbb{Z}_P^{\cdot} \to \pi_0(G_P/H) \to 0,$$

and so $\pi_0(G_p/H) = \mathbb{Z}_p^{\cdot +}$ (since the map $\mathbb{Z}/2\mathbb{Z} \to \mathbb{Z}_p^{\cdot}$ sends $-1 \mapsto -1$). Therefore $G_p/H \simeq (SG_p/SH) \times \mathbb{Z}_p^{\cdot +}$.

Break the fibration $BSH \to (BSG)_p$ up into the composition of $BSH \to (BSH)_p$ and $(BSH)_p \to (BSG)_p$. The fiber of the first is $(BSH)^{(P)}$, the colocalization, and the fiber of the second is $(SG/SH)_p \cong (G/H)_p$. Thus there is a fibration $(SG_p/SH) \to (G/H)_p$ with fiber $(BSH)^{(P)}$.

Since $\pi_i(G/H)$ and $\pi_i(BSH)$ are finite if $i \neq 4k$ and contain one factor of $\mathbb{Z}$ if $i=4k$,* the long exact sequence of the above fibration reduces to

$$0 \to \pi_i(BSH) \otimes \mathbb{Z}_{(P)} \to \pi_i(SG_p/SH) \to \pi_i(G/H) \otimes \mathbb{Z}_p \to 0$$

$$i = 4k+1, \quad 4k+2$$

$$0 \to \pi_{4k+3}((BSH)^{(P)})/(\mathbb{Z}_p/\mathbb{Z}) \to \pi_{4k+3}(SG_p/SH) \to \pi_{4k+3}(G/H) \otimes \mathbb{Z}_p \to 0$$

$$0 \to \pi_{4k}((BSH)^{(P)}) \to \pi_{4k}(SG_p/SH) \to \pi_{4k}(G/H) \otimes \mathbb{Z}_p \to \mathbb{Z}_p/\mathbb{Z} \to 0,$$

since any map $\pi_i(G/H) \otimes \mathbb{Z}_p \to \pi_{i-1}(BSH) \otimes \mathbb{Z}_{(P)}$ is necessarily 0 $(i \neq 4k)$.

This clearly implies the result for $i=4k+1$, $4k+2$. Since $\pi_{4k+3}((BSH)^{(P)}) \cong \pi_{4k+3}(BSH) \otimes \mathbb{Z}_{(P)} \oplus \mathbb{Z}_p/\mathbb{Z}$, it is also true for $i=4k+3$. Finally we have

$$0 \to \pi_{4k}((BSH)^{(P)}) \to \pi_{4k}(SG_p/SH)/\mathbb{Z} \to (\pi_{4k}(G/H)/\mathbb{Z}) \otimes \mathbb{Z}_p \to 0$$

which concludes the proof since

---

* See Theorem 4.4.3.

$$\pi_{4k}((BSH)^{(P)}) \oplus \mathbb{Z}_{(P)} \cong \pi_{4k}(BSH) \otimes \mathbb{Z}_{(P)}.$$

Note: In the case P = all primes, $G_P/H \simeq BSH \times \mathbb{Q}^{\cdot +}$.

We now show how the Spivak fibration and classifying space G_P/H apply to surgery theory, due in the case $P = \phi$ to Sullivan.

Let $(X, \partial X)$ be a Poincare pair over R of dimension n. An <u>H-normal map</u> is defined to be a degree 1 map

$\phi:(M,\partial M) \to (X,\partial X)$, where M is an H-manifold, together with an equivalence class of coverings of ϕ, $b:\mathcal{V}_M \to \xi$ for some H-bundle ξ over X , where $\mathcal{V}_M$ is the normal bundle of M in some large sphere. We assume $\phi|\partial M:\partial M \to \partial X$ is a homology equivalence over R .

Two normal maps (ϕ_1,b_1) , (ϕ_2,b_2) are <u>normally cobordant</u> if there is a map $\Phi:N \to X$ so that $\partial N = M_1 \cup M_2$, $M_1 \cap M_2 = \partial M_1 = \partial M_2$, $\Phi|M_i = \phi_i$, with an equivalence class of coverings $B:\mathcal{V}_N \to \xi$ so that $B|\mathcal{V}_{M_i}:\mathcal{V}_{M_i} \to \xi_i$ is equivalent to b_i .

The set of cobordism classes is denoted $NI^H(X;R)$ and is called the set of <u>normal invariants over R</u> of X . First consider the case $\partial X = \phi$.

<u>Theorem 3. Let $\mathcal{V}$ denote the Spivak fibration over $\mathbb{Z}_p$ of X . Then</u>

(1) $NI^H(X;\mathbb{Z}_p) \neq \phi$ <u>iff $\mathcal{V}$ is fiber homology equivalent over $\mathbb{Z}_p$ to an H-bundle over X ,</u>

(2) <u>and in this case,</u> $NI^H(X;\mathbb{Z}_p) \leftrightarrow [X,G_p/H]$.

Proof: (1) Let $\alpha \in H_{n+k}(T(\mathcal{V});\mathbb{Z}_p)$ denote a generator, and let ξ be an H-bundle over X and $h:\mathcal{V} \to \xi$ a fiber homology equivalence over $\mathbb{Z}_p$. The collapse defines a map $C:S^{n+k} \to T(\mathcal{V})$ so that $C_*(\iota_{n+k}) = \alpha$, where $\iota_{n+k} \in H_{n+k}(S^{n+k};\mathbb{Z}_p)$ is a generator. Making $T(h)C:S^{n+k} \to T(\xi)$ transverse regular to X , we get a normal map

$$\begin{array}{ccc} \nu_M & \longrightarrow & \xi \\ \downarrow & & \downarrow \\ M & \longrightarrow & X \end{array} \quad , \text{ where } M = (T(h)C)^{-1}X.$$

Conversely, let $\begin{array}{ccc} \nu_M & \xrightarrow{b} & \xi \\ \downarrow & & \downarrow \\ M & \xrightarrow{\phi} & X \end{array}$ be a normal map and let

$C_M:S^{n+k} \to T(\nu_M)$ be the collapse. Then ξ is a Spivak fibration over $\mathbb{Z}_p$ since

$$T(b)_*(C_M)_*:H_{n+k}(S^{n+k};\mathbb{Z}_p) \to H_{n+k}(T(\xi);\mathbb{Z}_p)$$

sends generator to generator.

(2) It is easy to see that $NI^H(X;\mathbb{Z}_p)$ is in 1-1 correspondence with the set of homotopy classes of lifts

$$\begin{array}{ccc} & & BH \\ & \overset{g}{\nearrow} & \downarrow \\ X & \xrightarrow[f]{} & (BG)_p \end{array}$$

where f denotes the classifying map for ν, and two lifts g_0,g_1 are homotopic if there exists a lift of $X \times I \underset{f}{\to} (BG)_p$, $G:X \times I \to BH$, so that $G|X \times i = g_i$, $i = 0,1$.

Let $E \to X$ denote the pullback of $BH \to (BG)_p$ by f. Then the fiber of $E \to X$ is G_p/H; let $T:(G_p/H) \times E \to E$ denote the action of the fiber on E. Clearly $NI^H(X;\mathbb{Z}_p)$

is the set of homotopy classes of sections of $E \to X$.

Since $NI^H(X;\mathbb{Z}_p) \neq \phi$, there is a section $s:X \to E$. Define $F:(G_p/H) \times X \to E$ by $F(y,x) = T(y,s(x))$. Then F is a homotopy equivalence and so sections of $E \to X$ correspond bijectively to maps $X \to G_p/H$.

If $\partial X \neq \phi$, then in order for $NI^H(X;\mathbb{Z}_p) \neq \phi$, we must have an H-manifold M and a normal homology equivalence over $\mathbb{Z}_p$, $M \to \partial X$. Thus we may assume $\mathcal{V}|\partial X$ is an H-bundle. Now we have $NI^H(X;\mathbb{Z}_p) \neq \phi$ iff $\mathcal{V}$ is fiber homology equivalent over $\mathbb{Z}_p$ rel $\mathcal{V}|\partial X$ to an H-bundle over X, and $NI^H(X;\mathbb{Z}_p) \longleftrightarrow$ the set of homotopy classes of maps $X \to G_p/H$ so that $\partial X \to X \to G_p/H$ maps to the basepoint.

If M is a manifold and $f:M \to X$ is a homology equivalence over R, then we will henceforth assume that f is a normal map. This is always the case if $R = \mathbb{Z}$.

Chapter 4. Surgery with Coefficients

4.1. Surgery.

Let M^n be a closed H-manifold and $f:S^k \times D^{n-k} \to M$ an embedding. Form $M' = (M-f(\mathrm{Int}(S^k \times D^{n-k}))) \bigcup_{f'} D^{k+1} \times S^{n-k-1}$, where $f' = f|S^k \times S^{n-k-1}$, straightening the angle if $H = \mathrm{DIFF}$. We say M' is obtained from M by surgery on the embedded sphere $f(S^k \times 0)$.

If $N = M \times I \bigcup_f D^{k+1} \times D^{n-k}$, where we regard $f:S^k \times D^{n-k} \to M \times 1$, then N is a cobordism from M to M', and is called the trace of the surgery. We say N is obtained from $M \times I$ by adding a $(k+1)$-handle.

Lemma 1. *Let N be a cobordism between manifolds M and M', with $\dim N \geq 6$ if $H = \mathrm{TOP}$. Then there is a sequence of embeddings $f_1,\dots,f_r$ so that N is H homeomorphic to the trace of the surgeries defined by $f_1,\dots,f_r$.*

Proof: This is the essence of Morse theory. See Milnor [D19] if $H = \mathrm{DIFF}$, Rourke and Sanderson [B12] if $H = \mathrm{PL}$, and Kirby and Siebenmann [J15] if $H = \mathrm{TOP}$. This trace is called a handlebody decomposition for N.

If $f:S^k \times D^{n-k} \to M$ is an embedding and N is the trace of the surgery defined by f, then $N \simeq M \bigcup_{f_0} D^{k+1}$, where $f_0 = f|S^k \times 0$. Let $\phi:M \to X$ be a map, where X is any space. It follows that ϕ admits an extension $\Phi:N \to X$ iff ϕf_0 is null-homotopic.

Define $\pi_{k+1}(\phi)$ to be the group of homotopy classes of commutative diagrams

$$\begin{array}{ccc} S^k & \subset & D^{k+1} \\ \downarrow & & \downarrow \\ M & \xrightarrow[\phi]{} & X \end{array} .$$

There is a long exact sequence

$$\cdots \to \pi_{k+1}(\phi) \to \pi_k(M) \xrightarrow{\phi_\#} \pi_k(X) \to \pi_k(\phi) \to \cdots .$$

Equivalently, $\pi_{k+1}(\phi) = \pi_{k+1}(X,M)$, where we replace ϕ by an inclusion. Clearly ϕ admits an extension iff f_0 comes from an element in $\pi_{k+1}(\phi)$.

If M has non-empty boundary, then we consider two cases, leaving the boundary fixed or doing surgery on the boundary. In the first case, assume $f:S^k \times D^{n-k} \to M$ is an embedding into the interior of M. Do surgery as before. Then the trace N is a manifold with $\partial N = M \cup (\partial M \times I) \cup M'$, $\partial M' = \partial M \times 1$. This is called <u>surgery relative to the boundary</u>.

If we wish to change the boundary, we need the following definition: $(M_1,\partial M_1)$ and $(M_2,\partial M_2)$ are <u>cobordant</u> if there is a manifold N with $\partial N = M_1 \cup P \cup M_2$ and $\partial P = \partial M_1 \cup \partial M_2$.

<u>Lemma 2</u>. <u>Any cobordism of</u> $(M,\partial M)$ <u>can be realized by attaching handles to</u> $\partial M \times I$ <u>followed by attaching handles to</u> $M \times I$ rel $\partial M \times I$.

Proof: Let N be a cobordism from $(M,\partial M)$ to $(M',\partial M')$, $\partial N = M \cup P \cup M'$, $\partial P = \partial M \cup \partial M'$. Define $Q = M \bigcup_{\partial M} P$; then $\partial Q = \partial M'$ and $Q \times I$ is a cobordism from $(M,\partial M)$ to $(Q,\partial Q)$ since

$$\begin{aligned}\partial(Q \times I) &= Q \times 0 \cup (\partial Q \times I) \cup Q \times 1 \\ &= M \times 0 \cup (P \cup \partial M' \times I) \cup Q \times 1.\end{aligned}$$

Also, N is a cobordism from $(Q,\partial Q)$ to $(M',\partial M')$ relative to the boundary since

$$\begin{aligned}\partial N &= M \cup P \cup M' \\ &= Q \cup M' \\ &= Q \cup (\partial Q \times I) \cup M'.\end{aligned}$$

By Lemma 1, $Q \times I \cup N$ is the desired cobordism.

4.2. The Problem of Surgery with Coefficients.

Let X be a finite CW-complex. The main problem considered here is the following: *When does* X *have the homology type over* R *of a manifold?*

To get a toe-hold on this problem, consider the following related problem: *if* M^n *is a manifold and* $\phi:M \to X$ *is a map, is* ϕ *cobordant to a homology equivalence over* R?

For simplicity, assume $R = \mathbb{Z}_p$. We thus wish to find a map $\phi':M' \to X$ cobordant to ϕ with $\pi_i(\phi') \otimes \mathbb{Z}_p = 0$. To do this, we would like to represent elements in $\pi_k(\phi)$ by embeddings $S^k \times D^{n-k} \to M$ and do surgery.

Let $f_0:S^k \to M$, be an embedding with ϕf_0 trivial. Since $S^k \times D^{n-k}$ is parallizable, if f_0 extends to $f:S^k \times D^{n-k} \to M$, then $f_0{}^*T_M$ must be trivial. We can do this if there is a bundle ξ over X so that $\phi^*\xi$ is the normal bundle of M in some large sphere. This is equivalent to saying that ϕ is a *normal map*, as in Section 3.2.

Lemma 1. Let $\phi:M^n \to X$ *be a normal map*, $n \geq 5$. *Then* ϕ *is normally cobordant to a map* $\phi':M' \to X$ *with* ϕ' $[n/2]$-*connected*.

The proof will be given below.

By Corollary 3.1.1, if X has the homology type over R of a manifold, then X must be a Poincare complex over R. Furthermore, a calculation shows that ϕ must have degree 1. Our problem is now reduced to two questions:

1) When is the Spivak fibration of X an H-bundle?
2) When is $\phi: M \to X$ normally cobordant to a homology equivalence over R?

Question 1 is best answered by analyzing the obstruction to lifting $X \to (BG)_p$ to BH. Question 2 will occupy the rest of these notes.

Proof of Lemma 1: Assume ϕ is an inclusion. Following Wall [H19] let $M = X_0, X_1, X_2, \ldots, X_m$ be a sequence of subcomplexes of X formed by attaching the cells of $X-M$ of dimension $\leq k = [n/2]$ one at a time. Assume by induction that there exist manifolds N_i, $N_0 = M \times I$, obtained by adding a handle of index $\leq k$ to N_{i-1} and a homotopy equivalence $\Phi_i: N_i \to X_i$ for $i \leq r$. We construct N_{r+1} and $\Phi_{r+1}: N_{r+1} \to X_{r+1}$.

Suppose $\partial N_r = M \cup M_r$, $\phi_r = \Phi_r | M_r$. The cell $X_{r+1} - X_r$ determines an element $\alpha \in \pi_i(X, X_r)$, where i is the dimension of that cell.

Since N_r is formed from $M \times I$ by attaching handles of index $\leq k$, it is formed from $M_r \times I$ by attaching handles of index $\geq (n+1-k) \geq k+1 \geq i+1$. Thus (N_r, M_r) is i-connected, and so $\pi_i(\phi_r) \cong \pi_i(X, X_r)$. Let $\alpha' \in \pi_i(\phi_r)$ correspond to α, and let

$$\begin{array}{ccc} S^{i-1} & \overset{j}{\subset} & D^i \\ {\scriptstyle f_0}\downarrow & & \downarrow{\scriptstyle g_0} \\ M & \xrightarrow[\phi]{} & X \end{array}$$

be a representative of α'.

Since ϕ is a normal map, there is a bundle ξ over X so that $T_M \oplus \phi^*\xi$ is stably trivial. Pulling back by f_0, $f_0^*T_M \oplus j^*g_0^*\xi$ is s-trivial. But D^i is contractible, so $j^*g_0^*\xi$ is trivial. Thus $f_0^*T_M$ is s-trivial. This defines a stable bundle monomorphism $T(S^{i-1} \times D^{n-i+1}) \to TM$. By the Immersion Classification Theorem, this defines an immersion $f:S^{i-1} \times D^{n-i+1} \to M$ so that $f|S^{i-1} \times 0 \simeq f_0$. Since $i-1 < [n/2]$, we may assume f is an embedding. Do surgery on f to get $\phi_{r+1}:N_{r+1} \to X_{r+1}$, a homotopy equivalence. Continuing inductively, we get $X_m = M \cup X^{(k)}$, $X^{(k)}$ the k-skeleton of X, so (X,X_m) is k-connected. Therefore (N_m,M_m) is k-connected and it follows that $\phi_m:M_m \to X$ is k-connected. It is easy to see that ϕ and ϕ_m are actually normally cobordant.

Corollary 1. *Let* $\phi:(M^n,\partial M) \to (Y,X)$ *be a normal map, where* (Y,X) *is a finite CW-pair*, X *a finite CW-complex. Then* ϕ *is normally cobordant to* $\phi':(M',\partial M') \to (Y,X)$ *where*

(a) *if* $n = 2k$, $\phi'|M'$ *is k-connected and* $\phi'|\partial M'$ *is (k-1)-connected*.

(b) *if* $n = 2k+1$, $\phi'|M'$ *and* $\phi'|\partial M'$ *are k-connected*.

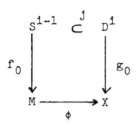

4.3. Surgery Obstruction Groups.

We describe here the functors of algebraic L-theory which provide us with surgery obstruction groups. Let Λ be a ring with involution $*$ and let I_k be the ideal generated by $x + (-1)^{k+1}x^*$, $x \in \Lambda$.

Definition. A $(-1)^k$-Hermitian form over Λ is a triple (G,λ,μ) where G is a free and based Λ-module, $\lambda : G \times G \to \Lambda$ and $\mu : G \to \Lambda/I_k$ are maps so that

(a) $\lambda(x,\ ) : G \to \Lambda$ is a λ-homomorphism

(b) $\lambda(y,x) = (-1)^k \lambda(x,y)^*$

(c) $\lambda(x,x) = \mu(x) + (-1)^k \mu(x)^*$

(d) $\lambda(x,y) = \mu(x+y) - \mu(x) - \mu(y)$

(e) $\mu(xa) = a^* \mu(x) a$.

(f) $A_\lambda : G \to G^*$, $A_\lambda(x)(y) = \lambda(x,y)$ is an isomorphism.

The torsion of the form (G,λ,μ) is defined to be the torsion of A_λ in $\overline{K}_1(\Lambda)$.

In case G has a basis e,f with $\mu(e) = \mu(f) = 0$, $\lambda(e,f) = 1$, then (G,λ,μ) is called a standard plane. A direct sum of standard planes is called a kernel.

Let $A \subset \overline{K}_1(\Lambda)$ be a self-dual subgroup. An $(-1)^k$-A-Hermitian form is one with torsion in A.

Definition. Let (G,λ,μ) be a $(-1)^k$ A-Hermitian form and $H \subset G$ a free and based submodule. Then H is called a subkernel of (G,λ,μ) if the basis of H extends to a preferred basis for G, making $^G/H$ based, and $\lambda(H \times H) = 0$, $\mu(H) = 0$, and the map $^G/H \to H^*$ defined by λ is an isomorphism with torsion in A, an A-isomorphism.

Lemma 1. (G,λ,μ) is a kernel iff there is a subkernel of (G,λ,μ).

Proof: If (G,λ,μ) is a kernel, then G has a basis $e_1,\ldots,e_n,f_1,\ldots,f_n$ so that $\mu(e_i) = \mu(f_i) = 0$, $\lambda(e_i,f_j) = \delta_{ij}$, $\lambda(e_i,e_j) = \lambda(f_i,f_j) = 0$. Let H be the submodule generated by $e_1,\ldots,e_n$. Then H is a subkernel.

Conversely, let $e_1,\ldots,e_n$ be the basis for H. Then, as $^G/H \to H^*$ is a A-isomorphism, we get a basis for $^G/H$, using the dual basis for H^*. Let $f_1',\ldots,f_n'$ be representatives of this basis in G.

Then $\lambda(e_i,f_j') = (-1)^k\delta_{ij}$, and $e_1,\ldots,e_n,f_1',\ldots,f_n'$ is a basis for G. Define

$$f_i = (-1)^k f_i' + (-1)^{k-1}(e_i\mu_i + \sum_{i<j} e_j\lambda(f_j',f_i')) \quad \text{where}$$

$\mu_i \in \mu(f_i')$.

Then $e_1,\ldots,e_n,f_1,\ldots,f_n$ is a basis for G and shows G is a kernel.

Corollary 1. Let H be a subkernel of (G,λ,μ) and H' a subkernel of (G',λ',μ'). Then an A-isomorphism $H \to H'$ extends to an A-isomorphism $(G,\lambda,\mu) \to (G',\lambda',\mu')$.

Definition. Define $L^A_{2k}(\Lambda)$ to be the semi-group of $(-1)^k A$-Hermitian forms over Λ, with addition defined by direct sum, modulo the equivalence relation $G \sim G'$ if there exist kernels K, K' so that $G \oplus K \cong G' \oplus K'$.

Lemma 2. $L^A_{2k}(\Lambda)$ is an abelian group.

Proof: All we need show is that $(G,\lambda,\mu) \oplus (G,-\lambda,-\mu)$ is a kernel. Let $e_1,\ldots,e_n$ be the basis of G. Denote by e_i', e_i'' the corresponding elements in $G \oplus G$, $e_i' = e_i + 0$, $e_i'' = 0 + e_i$.

Let $H \subset G \oplus G$ be generated by $e_i' + e_i''$. Then

$$\lambda(e_i' + e_i'', e_j' + e_j'') = 0, \quad \mu(e_i' + e_i'') = 0.$$

Furthermore, the map $G \oplus G/H \to H^*$ is given by

$$e_i' \mapsto \lambda(e_i', \quad)$$

and so has matrix (a_{ij}) where $a_{ij} = \lambda(e_i', e_j' + e_j'') = \lambda(e_i, e_j)$. But by hypothesis this matrix is in A. Thus H is a subkernel.

The group $L^A_{2k}(\Lambda)$ is called the (2k)-th Wall group of Λ.

To define odd dimensional surgery groups, let K_n denote the standard kernel with basis $e_1,\ldots,e_n,f_1,\ldots,f_n$, $\mu(e_i) = \mu(f_i) = 0$, $\lambda(e_i,e_j) = \lambda(f_i,f_j) = 0$, $\lambda(e_i,f_j) = \delta_{ij} = (-1)^k\lambda(f_j,e_i)$. Let $U_k(n,\Lambda)$ denote the group of isometries of K_n, i.e., the group of Λ-automorphisms $K_n \to K_n$ preserving λ and μ.

There is a map $U_k(n,\Lambda) \to \overline{K}_1(\Lambda)$ sending an isometry to its matrix; let $U_k^A(n,\Lambda)$ denote the inverse image of $A \subset \overline{K}_1(\Lambda)$.

Let $EU_k^A(u,\Lambda)$ be the subgroup of $U_k^A(n,\Lambda)$ generated by

(a) isometries which fix the subkernels generated by $\{e_1,\ldots,e_n\}$ and $\{f_1,\ldots,f_n\}$, and induce automorphisms with matrices in A.

(b) isometries which fix $e_1,\ldots,e_n$.

(c) the isometry σ_n defined by $\sigma_n(e_1) = f_1$, $\sigma_n(f_1) = (-1)^k e_1$, $\sigma_n(e_i) = e_i$, $\sigma_n(f_i) = f_i$, $i > 1$.

Then $EU_k^A(u,\Lambda)$ can be regarded as the group of matrices generated by matrices of the form

(a) $\begin{pmatrix} P & 0 \\ 0 & P^{*-1} \end{pmatrix}$, $P \in A' \subset GL(n,\Lambda)$, the inverse image of A,

(b) $\begin{pmatrix} I & Q \\ 0 & I \end{pmatrix}$, $Q = Q_0 + (-1)^{k+1} Q_0{}^*$ for some Q_0.

(c) the matrix $\left(\begin{array}{cc|c} 0 & 1 & \\ (-1)^k & 0 & 0 \\ \hline & & \\ 0 & & I \end{array}\right)$.

Embed $U_k^A(n,\Lambda)$ in $U_k^A(n+1,\Lambda)$ in the obvious way; then $EU_k^A(n,\Lambda) \subset EU_k^A(n+1,\Lambda)$.

Define $U_k{}^A(\Lambda) = \varinjlim U_k{}^A(n,\Lambda)$, $EU_k{}^A(\Lambda) = \varinjlim EU_k{}^A(n,\Lambda)$.

<u>The (2k+1)-th Wall group of</u> Λ is defined by

$$L^A{}_{2k+1}(\Lambda) = {}^{U_k{}^A(\Lambda)}/EU_k{}^A(\Lambda).$$

We will show in Corollary 5.2.1 that $L^A{}_{2k+1}(\Lambda)$ is an abelian group for $\Lambda = \mathbb{Z}_p\pi$. It is in fact abelian for all Λ, as is shown algebraically in Wall [W19].

Consider the following <u>surgery hypothesis</u>: Let $(M,\partial M)$ be a manifold pair with $\dim M = m \geq 5$, $(X,\partial X)$ a connected Poincare pair over $\mathbb{Z}_p$ of dimension n, $\phi:(M,\partial M) \to (X,\partial X)$ a normal map. Suppose that $A \subset Wh(\pi;\mathbb{Z}_p)$, $\pi = \pi_1(X)$, is self-dual and the torsion of $(X,\partial X)$ is in A; assume $\phi|\partial M:\partial M \to \partial X$ is a homology equivalence over $\mathbb{Z}_p$ with torsion in A, under the map $Wh(\pi_1(\partial X);\mathbb{Z}_p) \to Wh(\pi_1(X);\mathbb{Z}_p)$.

<u>Theorem 1</u>. <u>There is defined an element</u> $\sigma(\phi) \in L_n{}^A(\mathbb{Z}_p\pi)$, <u>which depends only on the normal cobordism class of</u> ϕ, <u>so that</u> $\sigma(\phi) = 0$ <u>if and only if</u> ϕ <u>is normally cobordant relative to the boundary to a homology equivalence over</u> $\mathbb{Z}_p$ <u>with torsion in</u> A.

Proof: <u>Case 1</u>. $n=2k$. By Lemma 4.2.1, we may assume ϕ is k-connected. By the Hurewicz theorem, $K_i(M) = 0$ for $i < k$ and $K_k(M) \cong \pi_{k+1}(\phi)$. By duality $K^i(M;\mathbb{Z}_p\pi) = 0$ for $i > k$, and so $K_i(M;\mathbb{Z}_p\pi) = 0$ for $i \neq k$. By Theorem 2.1, $G = K_k(M;\mathbb{Z}_p\pi) \cong \pi_{k+1}(\phi) \otimes \mathbb{Z}_p$ is s-free.

Then $G^* \cong K^k(M;\mathbb{Z}_p\pi)$ and Poincare duality gives an isomorphism $G \to G^*$ with matrix in A by the remarks following Theorem 3.1.2. This isomorphism defines $\lambda: G \times G \to \mathbb{Z}_p\pi$, which by Theorem 1.6.1 is given by intersection numbers, as every element of G is represented by an immersed k-sphere. Define $\mu: G \to \mathbb{Z}_p\pi/I_k$ by self-intersections.

If we do surgery on a trivial (k-1)-sphere in M, then we replace M by the connected sum $M\#(S^k \times S^k)$, and G is replaced by $G \oplus K_1$, $K_1 = \langle e,f\rangle$, where e is the class of $S^k \times 1$ and f is the class of $1 \times S^k$. Thus by adding trivial handles we may assume G is free. Routine calculations show that (G,λ,μ) is a $(-1)^k A$-Hermitian form. Define $\sigma(\phi)$ to be the class of (G,λ,μ) in $L^A_{2k}(\mathbb{Z}_p\pi)$.

To show $\sigma(\phi)$ depends only on the cobordism class of ϕ, let $\Phi: N \to X$ be a normal cobordism between ϕ and $\phi': M' \to X$, with $\partial N = M \cup \partial M \times I \cup M'$. Let $\psi: N \to I$ be an Urysohn function with $\psi(M) = 0$, $\psi(M') = 1$. Then the map

$$\Phi: (N;M,M') \to (X \times I; X \times 0, X \times 1)$$

sending x to $(\Phi(x),\psi(x))$ is a degree 1 normal map. By Corollary 4.2.1, we can assume $K_k(\Phi) = 0$, keeping M fixed and changing $K_k(M')$ by adding a kernel. The long exact sequence of $(N,\partial N)$ reduces to

$$0 \to K_{k+1}(N,\partial N;\mathbb{Z}_p\pi) \to K_k(\partial N;\mathbb{Z}_p\pi) \to K_k(N;\mathbb{Z}_p\pi) \to 0,$$

where we can assume all modules free and the torsion of the

complex above lies in A. Now, the form $K_k(\partial N;\mathbb{Z}_p\pi)$ represents $K_k(M;\mathbb{Z}_p\pi) - K_k(M';\mathbb{Z}_p\pi)$ in ${L^A}_{2k}(\mathbb{Z}_p\pi)$, and so we need only show that $K_{k+1}(N,\partial N;\mathbb{Z}_p\pi)$ is a subkernel. Duality shows that $K_{k+1}(N,\partial N;\mathbb{Z}_p\pi) \cong K^k(N;\mathbb{Z}_p\pi) \cong K_k(N;\mathbb{Z}_p\pi)^*$. Using $\Lambda = \mathbb{Z}_p\pi$ coefficients throughout, $K_k(\partial N)$ has a basis represented by framed immersions of spheres, so for $x \in K_{k+1}(N,\partial N)$, ∂x is represented by a sum of maps of spheres. This sum is 0 in $\pi_k(N)$, and so there is a framed immersion $\rho:(S^{k+1} - \bigcup \mathrm{Int}\, D^k) \to N$ so that ρ restricted to $\bigcup \partial D^k$ represents ∂x.

If ρ' represents $\partial x'$, then $\lambda(x,x') = \lambda(\partial x,\partial x') = 0$, since we can assume ρ and ρ' intersect transversally in a finite set of circles and paths with ends representing intersections of ∂x and $\partial x'$.([B12]) Similarly $\mu(\partial x) = 0$, and so $K_k(\partial N)$ is a kernel. Thus σ is a cobordism invariant.

Since if ϕ is a homology equivalence over $\mathbb{Z}_p$ with torsion in A, then $K_k(M;\mathbb{Z}_p) = 0$, we have $\sigma(\phi) = 0$ if ϕ is normally cobordant to a homology equivalence over $\mathbb{Z}_p$ with torsion in A.

Conversely, if $\sigma(\phi) = 0$, then $K_k(M;\mathbb{Z}_p)$ is a kernel, say $e_1,\ldots,e_m,f_1,\ldots,f_m$. Since $\mu(e_m) = 0$, e_m is represented by an embedding by Corollary 1.6.1. (This uses the fact that $\mathbb{Z}\pi \to \mathbb{Z}_p\pi$ is injective.) Do surgery on this embedding as in Lemma 4.2.1. Let M' be the result and N the trace of the surgery.

We have $N \simeq M \cup D^{k+1} \simeq M' \cup D^k$. The map

$j_*: K_k(M;\mathbb{Z}_p\pi) \to K_k(N,M';\mathbb{Z}_p\pi) \cong \mathbb{Z}_p\pi$ is easily seen to be given by $j_*(x) = \lambda(e_m, x)$. Thus j_* is surjective, since $\lambda(e_m, f_m) = 1$. Looking at the exact braid ($\mathbb{Z}_p\pi$-coefficients)

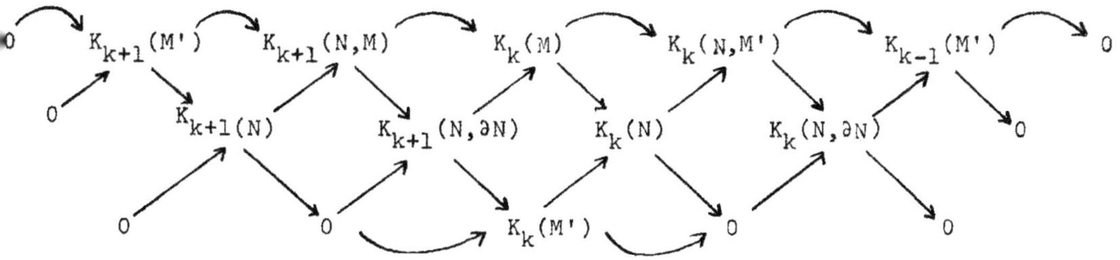

we see that $K_k(N,\partial N) = K_{k-1}(M') = 0$. Furthermore, $K_{k+1}(N,M) \cong \mathbb{Z}_p\pi$, since $N \simeq M \cup D^{k+1}$, and the map $\partial: K_{k+1}(N,M) \to K_k(M)$ sends the generator to e_m, since e_m represents the attaching map. Thus ∂ is injective and so $K_{k+1}(N) = K_{k+1}(M') = 0$.

A basis for $K_{k+1}(N,\partial N)$ is given by $e_1,\ldots,e_m,f_1,\ldots,f_{m-1}$, since $\ker(j_*)$ is generated by these elements. Also, the map $K_{k+1}(N,M) \to K_{k+1}(N,\partial N)$ sends the generator to e_m, and so $K_k(M') \cong \langle e_1,\ldots,e_{m-1},f_1,\ldots,f_{m-1}\rangle$. Inductively, we can do surgery to get a homology equivalence over $\mathbb{Z}_p$. Also, we chose bases so that $\tau(C_*(\phi)) \in A$ and took care that $\tau(C_*(\phi')) \in A$. Thus we have the result.

Case 2. $n = 2k+1$. Again assume ϕ is k-connected, so that $K_k(M) \cong \pi_{k+1}(\phi)$. Let $g_i: S^k \times D^{k+1} \to M$ be disjoint embeddings representing a set of generators of $\pi_{k+1}(\phi)$.

Let $U = \cup g_i(S^k \times D^{k+1})$, $M_0 = M - \text{Int } U$. By Theorem 3.1.3, we can assume there is a Poincare pair over $\mathbb{Z}_p$ with torsion in A, (X_0, S^{n-1}), so that $X = X_0 \cup D^n$. Changing ϕ by a homotopy, we can assume ϕ is a degree 1 map of triads $(M; M_0, U) \to (X; X_0, D^n)$.

The modules $K_{k+1}(U, \partial U; \mathbb{Z}_p\pi)$ and $K_k(U; \mathbb{Z}_p\pi)$ have bases given by $e_i = (g_i(1 \times D^{k+1}), g_i(1 \times S^k))$ and $f_i = g_i(S^k \times 1)$. Then in the sequence

$0 \to K_{k+1}(U, \partial U; \mathbb{Z}_p\pi) \to K_k(\partial U; \mathbb{Z}_p\pi) \to K_k(U; \mathbb{Z}_p\pi) \to 0$

these combine to give a basis $\{e_i, f_i\}$ for $K_k(\partial U; \mathbb{Z}_p\pi)$. In fact, $K_k(\partial U; \mathbb{Z}_p\pi)$ is a kernel and $K_{k+1}(U, \partial U; \mathbb{Z}_p\pi)$ is a subkernel. We have the diagram ($\mathbb{Z}_p\pi$ coefficients)

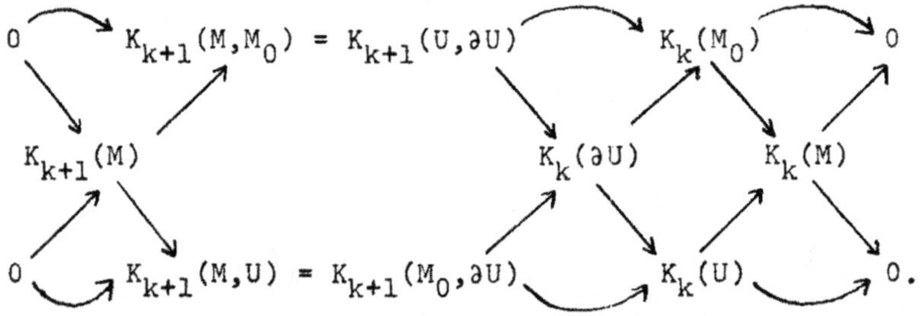

By theorem 2.1, $K_{k+1}(M_0, \partial U)$ is s-free and s-based. We can assume it is actually free. As above, $K_{k+1}(M_0, \partial U)$ is a subkernel. Let $\{e_i{}^*\}$ be a basis for $K_{k+1}(M_0, \partial U)$. Then the map $K_{k+1}(U, \partial U) \to K_{k+1}(M_0, \partial U)$, $e_i \mapsto e_i{}^*$, extends to an isometry $\alpha: K_k(\partial U) \to K_k(\partial U)$ by Corollary 4.3.1. Furthermore the matrix of α is in A. Define $\sigma(\phi)$ to be the class of α in $L^A{}_{2k+1}(\mathbb{Z}_p\pi)$.

To show σ is a cobordism invariant, note first that surgery on the embedding g_1 interchanges $g_1(S^k \times 1)$ and $g_1(1 \times S^k)$, and replaces α with $\alpha\sigma_n$. Thus $\sigma(\phi)$ is unaffected by surgeries on k-spheres. If $\psi: N \to X$ is a normal cobordism, then we can assume ψ is (k+1)-connected, and so the pairs $(N, \partial_+ N)$, $(N, \partial_- N)$ are k-connected. Thus N has a handle decomposition of handles of index >k. By Theorem 2.1, we can assume $K_{k+1}(N, \partial_+ N; \mathbb{Z}_p \pi)$ is free and based, with generators given by handles attached to $\partial_+ N$. Let $N' = \partial_+ N \times I \cup$ k-handles which generate $K_{k+1}(N, \partial_+ N; \mathbb{Z}_p \pi)$. Then $N = N' \cup N''$, where N'' is an s-cobordism over $\mathbb{Z}_p$ from $\partial_- N'$ to $\partial_- N$. Thus, over $\mathbb{Z}_p$, only k-handles need be considered, and so $\sigma(\phi)$ depends only on the cobordism class of ϕ.

Clearly $\sigma(\phi) = 0$ if ϕ is cobordant to a homology equivalence over $\mathbb{Z}_p$ with torsion A. Now suppose $\sigma(\phi) = 0$. Then $\alpha \in EU_k^A(r, \mathbb{Z}_p \pi)$. Write $\alpha = \alpha_0 \cdot \alpha_1 \ldots \alpha_m$,

where α_0 is of the form $\begin{pmatrix} R & 0 \\ 0 & R^{*-1} \end{pmatrix}$, and α_i is either of the form $\begin{pmatrix} I & Q \\ 0 & I \end{pmatrix}$ or (stably) $\begin{pmatrix} 0 & 1 \\ (-1)^k & 0 \end{pmatrix}$.

If $\alpha_m = \begin{pmatrix} 0 & 1 \\ (-1)^k & 0 \end{pmatrix}$, the surgery on the corresponding g_i replaces α with $\alpha\sigma$. But as $\sigma = (-1)^k \sigma^{-1}$, we may write $\alpha = \alpha_0 \ldots \alpha_{m-1}$.

If $\alpha_m = \begin{pmatrix} I & Q \\ 0 & I \end{pmatrix}$, then for some $p \in \Pi(P)$, pQ is a matrix over $\mathbb{Z}\pi$. Replacing g_i with $g_i \circ h$, where $h: S^k \times D^{k+1} \to S^k \times \text{Int } D^{k+1}$ is an embedding of degree p, replaces α with $\begin{pmatrix} pI & 0 \\ 0 & p^{-1}I \end{pmatrix} \alpha \begin{pmatrix} p^{-1}I & 0 \\ 0 & pI \end{pmatrix}$

$$= \begin{pmatrix} pR & \\ & (pR)^{*-1} \end{pmatrix} \alpha_1 \cdots \alpha_{m-1} \begin{pmatrix} p^{-1}I & 0 \\ 0 & pI \end{pmatrix} \begin{pmatrix} I & p^2Q \\ 0 & I \end{pmatrix}.$$

As $\begin{pmatrix} p^{-1}I & 0 \\ 0 & pI \end{pmatrix}$ is a simple isometry, we can assume $\alpha_m = \begin{pmatrix} I & Q \\ 0 & I \end{pmatrix}$ where Q is a matrix over $\mathbb{Z}\pi$ with $Q = Q_0 + (-1)^{k+1} Q_0^*$.

Let H be a regular homotopy of $\bigcup g_i : \bigcup S^k \times D^{k+1} \to M$ to disjoint embeddings, where the intersection matrix of the immersion H is given by $\begin{pmatrix} I & -Q \\ 0 & I \end{pmatrix}$ and self-intersection matrix given by $\begin{pmatrix} I & -Q_0 \\ 0 & I \end{pmatrix}$. This can be done similar to Theorem 1.6.3. The spheres $g_i(1 \times S^k)$ are unchanged, as they bound disks in M but the sphere $g_i(S^k \times 1)$ is replaced by the other end of the homotopy. This replaces α by

$$\alpha \cdot \begin{pmatrix} I & -Q \\ 0 & I \end{pmatrix} = \alpha_0 \cdots \alpha_{m-1}.$$

Thus by finitely many surgeries we can assume

$\alpha = \begin{pmatrix} R & 0 \\ 0 & R^{*-1} \end{pmatrix}$. But R has image in A, so by a change of basis, we can assume $\alpha=1$. By surgeries on the g_i, we can assume α is of the form $\pm\sigma \oplus \ldots \oplus \sigma$, i.e., $\alpha(e_i) = (-1)^k f_i$, $\alpha(f_i) = e_i$.

The bases of $K_{k+1}(M_0, \partial U)$ and $K_k(\partial U)$ determine a basis f_i^* for $K_k(M_0)$. Thus after all surgeries, the maps in the diagram above become ($\mathbb{Z}_p\pi$ coefficients):

$$K_k(\partial U) \to K_k(M_0): e_i \mapsto f_i^*, f_i \mapsto 0,$$

$$K_k(\partial U) \to K_k(U): f_i \mapsto f_i, e_i \mapsto 0,$$

$$K_{k+1}(M,U) \to K_k(\partial U): e_i^* \mapsto (-1)^k f_i,$$

$$K_{k+1}(M,U) \to K_k(U): e_i^* \mapsto (-1)^k f_i,$$

Thus the map $K_{k+1}(M,U) \to K_k(U)$ is an isomorphism with torsion in A, and so $K_{k+1}(M;\mathbb{Z}_p\pi) = K_k(M;\mathbb{Z}_p\pi) = 0$, with $\tau(C_*(\phi);\mathbb{Z}_p\pi) \in A$.

Theorem 2. (Realization Theorem). Let M^n be a connected, compact manifold, $n \geq 6$; let $\pi = \pi_1(M)$. Let $x \in L^A_{n+1}(\mathbb{Z}_p\pi)$. Then there is a manifold X and a normal map $\phi:(X;\partial_+X,\partial_-X) \to (M \times I; M \times 1, M \times 0 \cup \partial M \times I)$ so that $\phi|\partial_-X$ is the identity map, $\phi|\partial_+X$ is a homology equivalence over $\mathbb{Z}_p$ with torsion in A, and $\sigma(\phi) = x$.

Proof: Suppose $n+1 = 2k$. Then x is represented by a Hermitian form (G,λ,μ) with basis $e_1,\ldots,e_n$. Let $f_i:S^{k-1} \times D^k \to \operatorname{Int} M$ be embeddings into disjoint discs, $i=1,\ldots,r$. As in Theorem 1.6.3, let F_i be a regular homotopy to an embedding f_i' so that $\lambda(F_i,F_j) = p_{ij}\lambda(e_i,e_j)$ and $\mu(F_i) = p_i\mu(e_i)$, for some $p_{ij},p_i \in \Pi(P)$.

Let $X = M \times I \cup$ handles attached by the maps f_i'. Since the embeddings f_i were trivial, we can extend $1:M \to M$ to a normal map $\phi:(X;\partial_+X,\partial_-X) \to (M \times I; M \times 1, M \times 0 \cup \partial M \times I)$.

Then the intersection form on $K_k(X;\mathbb{Z}_p\pi)$ is isomorphic to (G,λ,μ), and the map $G \to G^*$ is given by $K_k(X;\mathbb{Z}_p\pi) \to K_k(X,\partial_+X;\mathbb{Z}_p\pi)$. Thus $K_k(\partial_+X;\mathbb{Z}_p\pi) = 0$. Since $k > 2$, $\pi_1(\partial_+X) \cong \pi$. Thus $\phi|\partial_+X$ is a homology equivalence over $\mathbb{Z}_p$ with torsion in A.

Now let $n+1 = 2k+1$, and let $\alpha:K \to K$ represent x, where K is the kernel of dimension $2r$. Do surgery on r trivial $(k-1)$-spheres in M. Let M' be the resultant and X' the trace; since the embedded spheres were trivial, we can again extend 1_M to $\phi':X' \to M \times I$.

Then $K_k(X';\mathbb{Z}_p\pi)$ is a kernel, with basis given by the $2r$ spheres $S^k \times 1, 1 \times S^k$. Let $v_1,\ldots,v_r$ be the images of $1 \times S^k$ under the map α. We can write $v_i = u_i \otimes a_i$ for $u_i \in K_k(X')$ and $a_i \in \mathbb{Z}_p$. The elements u_i are represented by disjoint framed embeddings; do surgery on these embeddings and let X'' be the trace. Then $X = X' \cup X''$ is the desired manifold.

4.4. The Simply Connected Case.

We assume $\pi = 1$ and compute $L_n(\mathbb{Z}_P)$. Torsion is not taken into account, as $Wh(1;\mathbb{Z}_P) = 0$.

Theorem 1.

$$L_n(\mathbb{Z}_P) = \begin{cases} 0 & n \text{ odd or } n=4k+2 \text{ and } 2 \in P \\ \mathbb{Z}/2\mathbb{Z} & n=4k+2 \text{ and } 2 \notin P \\ \mathbb{Z} \oplus \bigoplus_{p \in P} W_p^{\,P} & n=4k \end{cases}$$

where

$$W_p^{\,P} = \begin{cases} \mathbb{Z}/2\mathbb{Z} & p=2 \\ \mathbb{Z}/2\mathbb{Z} \oplus \mathbb{Z}/2\mathbb{Z} \text{ or } \mathbb{Z}/2\mathbb{Z} & p = 4k+1 \\ \mathbb{Z}/4\mathbb{Z} \text{ or } 0 & p = 4k+3 \end{cases}$$

Proof: Case 1. n odd. This proof follows Bernstein [G1]. Suppose A represents an automorphism of the standard kernel K_r. Assume inductively that all endomorphisms of K_{r-1} are equivalent mod $EU_{\pm 1}(r-1,\mathbb{Z}_P)$ to a matrix of the form

$$\begin{pmatrix} I & B_0 \\ 0 & C_0 \end{pmatrix}.$$

According to a matrix calculation given in [G1], Lemma 2.1, A is equivalent mod $EU_{\pm 1}(r,\mathbb{Z}_P)$ to a matrix $\begin{pmatrix} P & Q \\ R & S \end{pmatrix}$ with the element in the first row and column of P equal to 1, all other elements of P in the first row or first column 0, and the first row and column of R equal to 0.

Let A_0 be the matrix obtained by deleting the first rows and columns of P,Q,R, and S. By induction, we can assume $A_0 = \begin{pmatrix} I & B_0 \\ 0 & C_0 \end{pmatrix}$. Thus $U_{\pm 1}(r,\mathbb{Z}_p) = EU_{\pm 1}(r,\mathbb{Z}_p)$ and so $L_{2k+1}(\mathbb{Z}_p) = 0$.

<u>Case 2</u>. $n=4k$. Let $\mathbb{Q}_p$ denote the p-adic numbers for p a prime, and $\mathbb{F}_p$ the field of p elements. Then, since for F a field, $L_{4k}(F) = W(F)$, the Witt ring of F, for $p \neq 2$ there is a residue homomorphism $L_{4k}(\mathbb{Q}_p) \to L_{4k}(\mathbb{F}_p)$ (called the second residue in [A10], Theorem 1.6). Composing this with the functorial map obtained from $\mathbb{Z}_p \to \mathbb{Q}_p$, $p \in P$, we get $L_{4k}(\mathbb{Z}_p) \to L_{4k}(\mathbb{F}_p)$. Define $L_{4k}(\mathbb{Z}_p) \to \mathbb{Z}/2\mathbb{Z}$ if $2 \in P$ by sending $(G,\lambda,\mu) \mapsto \phi(\det \lambda) \bmod 2$, where ϕ is the 2-adic valuation.

Then there is an exact sequence

$$0 \to \mathbb{Z} \to L_{4k}(\mathbb{Q}) \to \mathbb{Z}/2\mathbb{Z} \oplus \bigoplus L_{4k}(\mathbb{F}_p) \to 0.$$

This is proved in Lam [A10], Theorem 4.1. The signature map $L_{4k}(\mathbb{Q}) \to \mathbb{Z}$ splits this sequence. The Chevalley-Warning Theorem ([A24]) and some elementary number theory show that the signature and ∂_p, $p \in P$, define an embedding of $W(\mathbb{Z}_p)$ in $\mathbb{Z} \oplus \bigoplus_{p \in P} L_{4k}(\mathbb{F}_p)$.

We let $\bar{W}_p^{\,P}$ = image of $\partial_p : L_{4k}(\mathbb{Z}_P) \to L_{4k}(\mathbb{F}_p)$, and $a_P \in \mathbb{Z}_+$ so that $a_P\mathbb{Z}$ = image of the signature $W(\mathbb{Z}_P) \to \mathbb{Z}$. We have

$$a_P = \begin{cases} 1 & \text{if } 2 \in P \\ 2 & \text{if } 2 \notin P\text{, but there is a prime of the form } 4k-1 \text{ in } P \\ 4 & \text{if neither hold, but there is a prime of the form } 4k+1 \text{ in } P \\ 8 & \text{if } P = \phi \end{cases}$$

and

$$\bar{W}_p^{\,P} = \begin{cases} \mathbb{Z}/2\mathbb{Z} & p=2 \\ \mathbb{Z}/4\mathbb{Z} & p\equiv 3 \mod (4) \\ \mathbb{Z}/2\mathbb{Z} \oplus \mathbb{Z}/2\mathbb{Z} & p\equiv 1 \mod (4) \end{cases}$$

The elements $\partial_p(G,\lambda,\mu) \in L_{4k}(\mathbb{F}_p)$ are called the <u>Hasse-Minkowski invariants</u> of the form (G,λ,μ). See Anderson, [K12], for proofs of these statements and a detailed study of Hasse-Minkowski invariants.

Case 3. $n=4k+2$. Suppose $2 \notin P$. Let (G,λ,μ) be a (-1)-Hermitian form, i.e., $\lambda: G \times G \to \mathbb{Z}_P$ non-degenerate anti-symmetric bilinear form and $\mu: G \to \mathbb{Z}_P/(2)$ a map so that $\mu(ax) = a^2\mu(x)$, $\lambda(x,y) = \mu(x+y) - \mu(x) - \mu(y)$, $\lambda(x,x) = 0$. Since $2 \notin P$, $\mathbb{Z}_P/(2) = \mathbb{Z}/2\mathbb{Z}$.

Apply the standard argument to get a sympletic basis for G, $e_1,\ldots,e_r,f_1,\ldots,f_r$ with $\lambda(e_i,e_j) = \lambda(f_i,f_j) = 0$ and $\lambda(e_i,f_j) = \delta_{ij}$. Define $c: L_{4k+2}(\mathbb{Z}_P) \to \mathbb{Z}/2\mathbb{Z}$ by the Arf invariant $c(G,\lambda,\mu) = \sum_{i=1}^{r} \mu(e_i)\mu(f_i)$.

Suppose $c(G,\lambda,\mu) = 0$. Then the number of i so that $\mu(e_i) = \mu(f_i) = 1$ is even. Group these in consecutive pairs. Apply the transformation

$$e_1' = e_1 + e_2 \qquad f_1' = f_1$$

$$e_2' = e_2 \qquad f_2' = f_1 + f_2,$$

and similarly obtain $e_2',\ldots,e_r',f_2',\ldots,f_r'$, leaving fixed those with $\mu(e_i) = 0$ or $\mu(f_i) = 0$.

Then $\langle e_1',f_2',\ldots,e_{r-1}',f_r'\rangle$ is a subkernel of G, so G is a kernel. The form $G = \langle e,f\rangle$, $\mu(e) = \mu(f) = 1$, $\lambda(e,e) = \lambda(f,f) = 0$, $\lambda(e,f) = 1$, has $c(G,\lambda,\mu) = 1$, so c is an isomorphism.

If $2 \in P$, then $\mathbb{Z}_P/(2) = 0$, and the same argument shows every such G is a kernel.

The groups $L_n(\mathbb{Z})$ were first considered in Kervaire-Milnor [F1]. See also Browder [G5]. We now construct the Milnor and Kervaire manifolds.

Let n be an integer, with $a_p | n$, and $n_p \in W_p^P$ for $p \in P$. Then $(n, (n_p)_{p \in P})$ determines an element $x \in L_{4k}(\mathbb{Z}_P)$, $k > 1$. By Theorem 3.3.2, there is a degree 1 map

$$\phi : (M; \partial_+ M, \partial_- M) \to (S^{4k-1} \times I, S^{4k-1} \times 1, S^{4k-1} \times 0)$$

so that $\sigma(\phi) = x$. Now $\partial_+ M \to S^{4k-1} \times 1$ is a homology equivalence over $\mathbb{Z}_P$ and so $\partial_+ M$ is a homology sphere over $\mathbb{Z}_P$. The cone of $\partial_+ M$, $C(\partial_+ M)$, is a Poincare complex over $\mathbb{Z}_P$ with boundary $\partial_+ M$. Define the <u>Milnor Poincare complex</u> by coning the boundary, $C(\partial_- M) \cup M \cup C(\partial_+ M)$. Define <u>Kervaire manifolds</u> similarly using $x \in L_{4k+2}(\mathbb{Z}_P)$.

<u>Theorem 2. There are Poincare complexes over</u> $\mathbb{Z}_P$, M^{4k}, $k > 1$, <u>and normal maps</u> $M \to S^{4k}$ <u>with signature</u> n <u>and Hasse-Minkowski invariants</u> n_p, <u>and</u> PL-<u>manifolds</u> K^{4k+2}, $k > 0$, <u>and normal maps</u> $K \to S^{4k+2}$ <u>with Arf invariant</u> 1.

Theorem 3. Let H be PL or TOP. Then $\pi_n(G/H) \cong L_n(\mathbb{Z})$, $n \geq 5$.

Proof: By Theorem 3.2.3, $\pi_n(G/H) \leftrightarrow NI^H(S^n)$. But this is generated by the Milnor map $M^{4k} \to S^{4k}$ if $n=4k$ and the Kervaire map $K^{4k+2} \to S^{4k+2}$ if $n=4k+2$, and is 0 if n is odd.

4.5. The Exact Sequence of Surgery.

An important corollary of the surgery theorems of Section 4.3 is the exact sequence of surgery. In the closed, simply connected case, $P = \phi$, it is due to Sullivan [G13]; the general case, $P = \phi$, is due to Wall [H19].

Let $\mathcal{N}_P{}^H(X)$ be the set of homology equivalences over $\mathbf{Z}_P$, $f:M \to X$, with torsion in $A \subset Wh(\pi;\mathbf{Z}_P)$, $\pi = \pi_1(X)$, where M is an H-manifold, modulo the relation $f_0 \sim f_1$ if there is an h-cobordism over $\mathbf{Z}_P$, W, with torsion in A, between M_0 and M_1, and a map $F:W \to X$ extending f_0 and f_1. Here X is a Poincare complex over $\mathbf{Z}_P$ with torsion in A, and all maps are assumed to be normal. The subgroup A is suppressed from the notation.

Theorem 1. There is an exact sequence

$$L_{n+1}(\mathbf{Z}_P\pi) \xrightarrow{\omega} \mathcal{N}_P{}^H(X) \xrightarrow{\alpha} [X,G_P/H] \xrightarrow{\sigma} L_n(\mathbf{Z}_P\pi)$$

provided X is an H-manifold of dimension $n \geq 5$.

Proof: The map $\mathcal{N}_P{}^H(X) \to [X,G_P/H]$ is defined by Theorem 3.2.3, as elements of $\mathcal{N}_P{}^H(X)$ are represented by normal maps. The map (not a homomorphism in general) σ is defined by taking surgery obstructions. To define ω, let $x \in L_{n+1}(\mathbf{Z}_P\pi)$. Then there is a normal map $\phi:(M;\partial_+M,\partial_-M) \to (X \times I;X \times 1,X \times 0)$ with $\sigma(\phi) = x$ and $\phi|\partial_+M$ a homology equivalence over $\mathbf{Z}_P$ with torsion in A.

Define $\omega(x)$ = the class of $\phi|\partial_+ M$ in $\mathcal{N}_p{}^H(X)$.

In fact, this procedure defines an action of $L_{n+1}(\mathbb{Z}_p\pi)$ on $\mathcal{N}_p{}^H(X)$, by taking a homology equivalence over $\mathbb{Z}_p$, $f:N \to X$, and doing as above to N, getting a homology equivalence $\partial_+ M \to N$. Composition defines the action.

Exactness in the theorem then means that α induces a bijection of the orbits of the action to the kernel of σ. This is the content of Theorems 4.3.1 and 4.3.2.

Chapter 5. Relative Surgery.

5.1. Handle Subtraction and Applications.

In this section we use handle subtraction, an operation dual to surgery, to prove a general relative surgery lemma which forms the basis for the geometric formulation of surgery groups.

Let $\Psi:(N,M) \to (Y,X)$ be a map of pairs and define $\pi_r(\Psi)$ to be the set of homotopy classes of diagrams

$$\begin{array}{ccc} (D_-^{r-1},S^{r-2}) & \to & (D^r,D_+^{r-1}) \\ \downarrow & & \downarrow \\ (N,M) & \to & (Y,X). \end{array}$$

If N^n is a manifold with boundary M and ψ is a normal map, then each $\alpha \in \pi_{r+1}(\psi)$ determines a regular homotopy class of immersions $f:(D_-^r x D^{n-r}, S^{r-1} x D^{n-r}) \to (N,M)$ for $r \leq n-2$, by the relative immersion classification theorem ([B3]). If this class contains an embedding, let

$$N_0 = N - \text{Int } f(D_-^r x D^{n-r})$$

$$M_0 = \partial N_0.$$

Since $\alpha \in \pi_{r+1}(\psi)$, ψ induces $\psi_0:(N_0,M_0) \to (Y,X)$. Furthermore $\pi_{r+1}(\psi_0) = \pi_{r+1}(\psi)/\langle\alpha\rangle$. N and N_0 are cobordant by attaching an $(n-r)$-handle to $N_0 xI$.

Theorem 1. Let $\phi:(N^n;M,M_+) \to (Y;X,X_+)$ be a normal map, where $(N;M,M_+)$ is a manifold triad, $(Y;X,X_+)$ is a Poincare triad over $\mathbb{Z}_P$ with torsion in $A \subset Wh(\pi;\mathbb{Z}_P)$, $\pi = \pi_1(X) \cong$

$\pi_1(Y)$, induced by inclusion, $\phi|M_+$ a homology equivalence over $\mathbb{Z}_p$ with torsion in A, and $n \geq 6$. Then ϕ is cobordant rel M_+ to a homology equivalence over $\mathbb{Z}_p$ with torsion in A.

Proof: Even-dimensional case, $n = 2k$.

By Corollary 4.2.1., we may assume $\phi|N$ is k-connected and $\phi|M$ is (k-1)-connected. By duality, $K_i(N,M) = 0$ for $i \neq k$ and we use $\mathbb{Z}_p\pi$ coefficients unless otherwise noted. By Theorem 2.1, $K_k(N,M)$ is s-free. By adding trivial handles, we can assume it is free with basis $e_1,\ldots,e_r$.

By the Hurewicz theorem, $K_k(N,M) \cong \pi_{k+1}(\phi) \otimes \mathbb{Z}_p$, where we regard $\phi:(N,M) \to (Y,X)$. Thus the elements e_i determine immersions $f_i':(D^k x D^k, S^{k-1} x D^k) \to (N,M)$; f_i' represents $qe_i \in \pi_{k+1}(\phi)$ for some $q \in \Pi(P)$. Since $\pi_1(M) \cong \pi_1(N)$, the maps f_i' are regularly homotopic to disjoint embeddings f_i, by Corollary 1.6.2.

Do handle subtraction: Let $N_0 = N - \text{Int } U$, $M_0 = \partial N_0$, where $U = \bigcup f_i(D^k x D^k)$. Let $C_*(\phi)$ be the chain complex of ϕ, given by the chain complex of $(Y, N \cup X)$ if ϕ is replaced by an inclusion, and $H_*(\phi) = H(C_*(\phi))$ (tensoring as in Section 1.2 if using coefficients). For any coefficients there is an exact sequence

$$\cdots \to H_i(N,M) \xrightarrow{\phi_*} H_i(Y,X) \to H_i(\phi) \to H_{i-1}(N,M) \to \cdots.$$

By Theorem 3.1.2, $H_i(\phi) = K_{i-1}(N,M)$. Let D_* be the chain complex defined by $D_i = C_{i+1}(\phi) \otimes \mathbb{Z}_p\pi$. We have $H_k(U \cup M, M) \xrightarrow{\cong} K_k(N,M)$, and it follows that

$C_*(U \cup M, M) \otimes \mathbb{Z}_P\pi \to D_*$ is a chain equivalence with torsion in $\dot{A}$. As $(N_0, M_0) \to (N, U \cup M)$ is an excision, we have a chain equivalence $C_*(N_0, M_0) \otimes \mathbb{Z}_P\pi \to C_*(Y, X) \otimes \mathbb{Z}_P\pi$ with torsion in A. By Poincare duality, $C_*(N_0) \otimes \mathbb{Z}_P\pi \to C_*(Y) \otimes \mathbb{Z}_P\pi$ is a chain equivalence and it follows that $C_*(M_0) \otimes \mathbb{Z}_P\pi \to C_*(X) \otimes \mathbb{Z}_P\pi$ is also. Thus $\phi_0 : (N_0, M_0) \to (Y, X)$ is a homology equivalence over $\mathbb{Z}_P$ with torsion in A.

<u>Odd-dimensional case, $n = 2k+1$</u>.

We can regard ϕ as a normal cobordism from $\phi | M$ to $\phi | M_+$, and so $\sigma(\phi | M) = 0$, since $\phi | M_+$ is a homology equivalence over $\mathbb{Z}_P$ with torsion in A. Here σ denotes the surgery obstruction in $L^A_{n-1}(\mathbb{Z}_P\pi)$. Let $\Phi : Q \to X \times I$ be a normal cobordism to a homology equivalence over $\mathbb{Z}_P$ with torsion in A. Then $\phi \cup \Phi : N \cup Q \to Y \cup X \times I$ has a well-defined obstruction $x \in L^A_n(\mathbb{Z}_P\pi)$.

Let $\psi : R \to X \times I$ be a normal map with $\sigma(\psi) = -x$. Then $\sigma(\phi \cup \Phi \cup \psi) = x - x = 0$, and so $\phi \cup \Phi \cup \psi : N \cup Q \cup R \to Y \cup X \times I$ is cobordant, relative to the boundary, to a homology equivalence over $\mathbb{Z}_P$ with torsion in A. This can be regarded as a cobordism of (N, M) rel M_+ to a homology equivalence over $\mathbb{Z}_P$ with torsion in A.

For the case $P = \emptyset$, this theorem is due to Wall, [G18] for the case $\pi = 1$, and [H19] for the general case. The proof of the odd-dimensional case is due to Cappell and Shaneson [K5].

5.2. Geometric Definitions of Surgery Groups.

In this section we define surgery groups in a more general context and relate them to the algebraic definitions given in Section 4.3. To do this we need the notion of an n-ad.

Let $\mathcal{C}$ be a category of spaces and maps and $n \geq 2$ an integer. Define $\mathcal{C}^{(n)}$ to be the category with objects

$$X = (|X|; X_1, \ldots, X_{n-1})$$

$X_i \subset |X|$, and $X(\alpha) = \bigcap_{i \notin \alpha} X_i$, $\alpha \subset \{1, \ldots, n-1\}$, is an object of $\mathcal{C}$

and morphisms between X and Y given by a map $f: |X| \to |Y|$ so that $f(X(\alpha)) \subset Y(\alpha)$ for each α. We let $X\{1, \ldots, n-1\} = |X|$.) This is called the category of n-ads associated to $\mathcal{C}$.

Define functors ∂_i, δ_i, and $s_{n,k}$ by:

$\partial_i : \mathcal{C}^{(n+1)} \to \mathcal{C}^{(n)}$,

$$|\partial_i X| = X_i, \quad (\partial_i X)_j = \begin{cases} X_i \cap X_j & j < i \\ X_i \cap X_{j+1} & j \geq i, \end{cases}$$

$\delta_i : \mathcal{C}^{(n+1)} \to \mathcal{C}^{(n)}$,

$$|\delta_i X| = |X|, \quad (\delta_i X)_j = \begin{cases} X_j & j < i \\ X_{j+1} & j \geq i, \end{cases}$$

$s_{n,k} : \mathcal{C}^{(n)} \to \mathcal{C}^{(n+k)}$,

$$|s_{n,k} X| = |X|, \quad (s_{n,k} X)_j = \begin{cases} X_j & j < n \\ X & j \geq n. \end{cases}$$

Define $\partial X(\alpha) = \bigcup_{\beta \subsetneq \alpha} X(\beta)$. In particular, $\partial X = \bigcup_{i=1}^{n-1} X_i$.

If $f:X \to Y$ is a map of n-ads, let $\partial_i f:\partial_i X \to \partial_i Y$ be the induced map; similarly for δ_i, ∂ and $s_{n,k}$. If X is an n-ad and Y is a space, define an n-ad XxY by $|X \times Y| = |X| \times Y$, $(X \times Y)_i = X_i \times Y$.

Let X be an n-ad with $x_o \in X(\emptyset)$. Define an (n-1)-ad X' by $|X'| = [|X|;X_1,x_o]$, $X'_i = [X_{i+1};X_1 \cap X_{i+1},x_o]$, where $[C;A,B] = \{\sigma \in C^I:\sigma(0) \in A, \sigma(1) \in B\}$ for $A,B \subset C$. Define inductively $\pi_m(X,x_o) = \pi_{m-1}(X',x'_o)$ where x'_o is the constant path at x_o and $m \geq n - 1$.

An n-ad M is a manifold n-ad if each $M(\alpha)$ is a manifold with boundary $\partial M(\alpha)$. An n-ad X is a Poincare n-ad (over a ring) if for each $\alpha = \{i_1,\ldots,i_k\}$, $(X(\alpha),\partial X(\alpha))$ is a Poincare pair with fundamental class $[X(\alpha),\partial X(\alpha)]$ so that

$$\partial[X(\alpha),\partial X(\alpha)] = j_* \sum_{t=1}^{k} (-1)^t [X(\alpha - i_t)], \text{ where}$$

$$j: \bigcup_{t=1}^{k} X(\alpha - i_t) \hookrightarrow \partial X(\alpha).$$

A map $\phi:X \to Y$ between Poincare n-ads is of degree 1 if $\phi_*[X(\alpha),\partial X(\alpha)] = [Y(\alpha),\partial Y(\alpha)]$ for each α.

Let $\overline{K}$ be a pair (K,w), where K is a CW (n-1)-ad and $w \in H^1(|K|; \mathbb{Z}/2\mathbb{Z})$. Let $\mathcal{C}$ be a subcategory of the category of normal maps of degree 1, $\phi:M \to X$, between an H-manifold n-ad M and a Poincare n-ad X over a ring R, closed under the functors ∂, ∂_i. Here H = TOP, PL, or DIFF, and we omit further mention of it.

Define $\Omega_m^{\mathcal{C}}(\overline{K})$ to be the cobordism group of $\mathcal{C}$ over $\overline{K}$, where we regard a map $(\phi:M \to X) \to \overline{K}$ as a map of n-ads $\omega:X \to s_{n-1,1}K$ so that $w_{|X|} = \omega^* w$, where $w_{|X|}$ is the orientation class of $|X|$, and we use the boundary operator $\partial_{n-1} + \partial_n$. Thus $M_1 \to X_1 \to \overline{K}$ and $M_2 \to X_2 \to \overline{K}$ are bordant if there are maps of (n+1)-ads $N \to Y \to s_{n-1,2}K$ as above so that $\partial_n N \to \partial_n Y \to \partial_n s_{n-1,2}K = s_{n-1,1}K$ is equal to $M_1 \to X_1 \to s_{n-1,1}K$ and similarly applying ∂_{n-1} yields $M_2 \to X_2 \to \overline{K}$. Note we also require compatible orientation relation for Y as above. $\Omega_m(\overline{K})$ denotes the group defined above with $\mathcal{C}$ the full category. The integer m denotes the dimension of $|M|$.

Let $\Pi = (\overline{K},R,A)$ where R is a ring and A is a self-dual subgroup of $Wh(\pi_1(|K|);R)$. Define $Q_m(\Pi) = \Omega_m^{\mathcal{C}}(\bar{K})$ where $\mathcal{C}$ is the subcategory of homology equivalences over R with torsion in A. There is a natural map $h_\Pi:Q_m(\Pi) \to \Omega_m(\bar{K})$.

Define $\Omega_m(\Pi) = \partial_{n-1}^{-1} h_\Pi Q_{m-1}(\Pi)$, where $\partial_{n-1}:\Omega_m(\bar{K}) \to \Omega_{m-1}(\bar{K})$, and $\mathcal{R}_m(\Pi) = h_\Pi^{-1}\partial_{n-1}^{-1} h_\Pi Q_{m-1}(\Pi)$. Define $\Omega_m'(\Pi)$ and $\mathcal{R}_m'(\Pi)$ similarly by requiring that $\omega:\delta_{n-1}X \to K$ induce isomorphisms on fundamental groupoids (that is, for each α the corresponding intersections have isomorphic fundamental groups on each component. This will be made clear later.); we also assume X is connected. and we require the same for cobordisms.

There is a natural map $\mathcal{R}_m(\Pi) \to \Omega_m(\Pi)$ and we define $L_m(\Pi)$ to be $\Omega_m(\Pi)$/image of $\mathcal{R}_m(\Pi)$; define $L_m'(\Pi)$ similarly. There is a natural map $L_m'(\Pi) \to L_m(\Pi)$; however $L_m'(\Pi)$ has no natural group structure.

The value of the restricted set $L'_m(\Pi)$ is shown in the next theorem. Suppose $\phi: M \to X$ is a normal map between a manifold n-ad and a Poincare n-ad over $\mathbb{Z}_p$ with torsion in A, $\partial_{n-1}\phi$ a homology equivalence over $\mathbb{Z}_p$ with torsion in A. Let K be an (n-1)-ad of type $(\pi_1(\delta_{n-1}X),1)$, that is an (n-1)-ad so that the total space and each subspace are $K(\pi,1)$'s corresponding to the fundamental groups of the spaces in $\delta_{n-1}X$ (and consistent with the inclusion maps). Then $w_{|X|}$ is classified by a map $X \to s_{n-1,1}K$, and defines an orientation class w in $H^1(|K|;\mathbf{Z}/2\mathbf{Z})$. Let $\Pi = (K, w, \mathbb{Z}_p, A)$.

<u>Theorem 1. With notation as above, if dim M = m, and $m - n \geq 3$, then ϕ is normally cobordant to a homology equivalence of n-ads over $\mathbb{Z}_p$ with torsion in A if and only if the class of $M \to X \to K$ in $L'_m(\Pi)$ vanishes.</u>

Proof: Suppose first that n = 2. Assume ϕ is cobordant to a homology equivalence over $\mathbb{Z}_p$, $\phi_+: M_+ \to X$, by a cobordism $\psi': N \to X$, with $\partial N = M \cup M_+$, $M \cap M_+ = \partial M = \partial M_+$.

Let $Y = X \times I/\sim$ where $(x,0) \sim (x,t)$ for $x \in \partial X$, $t \in I$. Let $f: N - \partial M \to I$ be a Urysohn function with $f(M) = 0$, $f(M_+) = 1$. Define $\psi: N \to Y$ by

$$\psi(x) = \begin{cases} (\psi'(x), f(x)) & x \notin \partial M \\ (\psi'(x), 0) & x \in \partial M. \end{cases}$$

Then $\psi: (N; M, M_+) \to (Y; X\times 0, X\times 1)$ is a normal map and shows that the class of $M \to X$ vanishes is $L'_m(\Pi)$.

Now suppose the class of $M \to X \to K$ vanishes. Then there is a cobordism $\psi: (N; M, M_+) \to (Y; X, X_+)$ to a homology equivalence $\phi_+:M_+ \to X_+$ over $\mathbb{Z}_P$ (with torsion in A). Furthermore, $\pi_1(X) \cong \pi_1(K) \cong \pi_1(Y)$, and it follows easily that $\pi_1(X) \cong \pi_1(Y)$, induced by inclusion.

By Theorem 5.1.1, we can do surgery on ψ to get a homology equivalence over $\mathbb{Z}_P$. In particular, we can do surgery on $\phi:M \to X$.

The remainder of the theorem follows by induction: suppose surgery has been done on $M(\beta)$ for each $\beta \subsetneq \alpha$. Then apply the procedure above to the pair $(M(\alpha),\partial M(\alpha))$.

Thus $L'_m(\Pi)$ is an obstruction theory for surgery on n-ads. We now show that $L'_m(\Pi)$ has nice group and functorial properties by proving $L'_m(\Pi) \to L_m(\Pi)$ is a bijection.

Lemma 1. Let $\phi:(W,V) \to (Y,X)$ be a normal map, (Y,X) a Poincare pair over $\mathbb{Z}_P$ with torsion in A, $\dim W = m \geq 5$, and $\omega:Y \to K$ so that $w_Y = w\circ\omega_\#:\pi_1(Y) \to Z/2Z$. Assume K has a finite 2-skeleton. Then there is a normal cobordism of ϕ, $\psi:(U;W,W_+) \to (Z;Y,Y_+)$, $W \cap W_+ = V$, $Y \cap Y_+ = X$, and an extension of ω, $\Omega:Z \to K$ so that $(\Omega|Y_+)_\#:\pi_1(Y_+) \to \pi_1(K)$ is an isomorphism.

Proof: By Theorem 3.1.3, we can assume $Y = Y_0 \underset{\partial H}{\cup} H$, $X \subset Y_0$, $\dim(Y_0 - X) \leq m - 2$, and H is obtained from D^m by adding 1-handles. The inclusion induces a surjection $\pi_1(H) \to \pi_1(Y)$.

If ξ is the line bundle over K defined by w, then $TH + (\omega|H)^*\xi$ is trivial, so we can do surgery on $\omega|H:H \to K$

to get $H' \to K$ which induces an isomorphism on fundamental groups.

Let J be the trace of the surgery, and define $Z_0 = Y_0 \times I \cup J$, $Y_+ = Y_0 \cup H'$, and $Z = Z_0/\sim$, where $(x,t) \sim (x,0)$ for $x \in X$, $t \in I$. Since $\omega|H$ extends over J, we get a map $\Omega : Z \to K$.

By construction, $\pi_1(H') \to \pi_1(Y_+) \xrightarrow{\Omega_\#} \pi_1(K)$ is an isomorphism; thus $(\Omega|Y_+)_\#$ is onto. But $\pi_1(H') \to \pi_1(Y_+)$ is also onto and so $(\Omega|Y_+)_\#$ is an isomorphism. To construct U, we consider the handles of J, taking one at a time.

<u>1-handle case</u>. Assume ϕ is transverse regular to the embedded sphere S^0 and to H. Let $S = \phi^{-1}(S^0)$, $T = \phi^{-1}(H)$. Then $\phi|T : T \to H$ has degree $r \in \Pi(P)$ and so the total multiplicity of each component of S^0 is r.

Write $S^0 = \{a,b\}$ and choose $x_1,\ldots,x_r \in \phi^{-1}(a)$, $y_1,\ldots,y_r \in \phi^{-1}(b)$ with multiplicity 1. Attach handles to discs containing x_i, y_i and extend ϕ. We can now arrange the other points of S in complementary pairs having the same image but opposite degrees. Add a handle along each such pair and extend ϕ.

<u>2-handle case</u>. Let $S = \phi^{-1}(S^1)$ (assuming ϕ is transverse regular). We can assume W is connected, and so by joining components by paths, we can assume S is connected. Then $\phi|S : S \to S^1$ is of degree $r \in \Pi(P)$ and so is homotopic to an embedding $S^1 \times D^{m-1} \to S^1 \times \mathrm{Int}(D^{m-1})$ of degree r. Add a handle to $W \times I$, which gives the result.

Theorem 2. If $m - n \geq 3$ and $|K|$ has a finite 2-skeleton, then $L'_m(\Pi) \to L_m(\Pi)$ is a bijection.

Proof: Follows immediately from the lemma.

Corollary 1. $L^A_m(K,w;\mathbb{Z}_P)$ is isomorphic to $L^A_m(\mathbb{Z}_P\pi_1(K))$ if $m \geq 5$.

It follows that $L^A_{2k+1}(\mathbb{Z}_P\pi)$ is abelian, since it is isomorphic to the geometrically defined group, which is abelian. Note also that though $\mathcal{R}_n(\Pi)$ and $\Omega_n(\Pi)$ depend on H = TOP, PL, or DIFF, $L_m(\Pi)$ is independent of H. Also, by Theorem 4.3.2, we need only consider normal maps with the target space a manifold.

Lemma 2. Let R be a principal ring, and $P = \{p : p \text{ a prime and } R \otimes \mathbb{Z}/p\mathbb{Z} = 0\}$. If C is a chain complex of free $\mathbb{Z}\pi$ modules, then $C \otimes \mathbb{Z}_P\pi$ is acyclic if and only if $C \otimes R\pi$ is acyclic.

Proof: Since $C \otimes R\pi \cong (C \otimes R) \otimes \mathbb{Z}\pi$, we need only show that $C \otimes R$ is acyclic, and similarly for $\mathbb{Z}_P$.

Now suppose $C \otimes \mathbb{Z}_P$ is acyclic. Since $\mathbb{Z}_P$ is torsion free, $0 = H_i(C \otimes \mathbb{Z}_P) \cong H_i(C) \otimes \mathbb{Z}_P$. Thus $H_i(C) \otimes R = 0$. So we have

$$\begin{aligned} H_i(C \otimes R) &\cong H_i(C) \otimes R \oplus H_{i-1}(C) * R \\ &= H_{i-1}(C) * R. \end{aligned}$$

Thus the proof is reduced to showing $A \otimes R = 0$ implies $A * R = 0$ for A a finitely generated abelian group. Since $\otimes$ and $*$ commute with direct sum, we can assume $A = \mathbb{Z}/n\mathbb{Z}$. The exact sequence

$$0 \to \mathbb{Z} \xrightarrow{n} \mathbb{Z} \to A \to 0$$

is a free presentation for A so we have

$$0 \to A*R \to R \xrightarrow{n} R \to A \otimes R \to 0.$$

Thus $A*R \cong R/nR$

$\cong A \otimes R.$

Conversely, if $C \otimes R$ is acyclic, $H_i(C) \otimes R = 0$, and so $H_i(C) \otimes \mathbb{Z}_p = 0$. So $C \otimes \mathbb{Z}_p$ is acyclic.

So if A is a self-dual subgroup of $Wh(\pi;R)$, let $A' = \Phi_*^{-1}A$, where $\Phi_*: Wh(\pi;\mathbb{Z}_p) \to Wh(\pi;R)$ is induced by the unique ring homomorphism $\Phi: \mathbb{Z}_p \to R$.

Define $L_m^A(\pi,w;R) = L_m^{A'}(K(\pi,1),w;\mathbb{Z}_p)$. **We let** L_m^h, L_m^s **denote** L_m^A **for** $A = Wh(\pi;R)$, 0.

We now clarify the algebraic nonsense of n-ads as promised earlier. Let $\mathcal{C}$ be the category of finitely generated groupoids. A <u>groupoid of type 2^n</u> is an object π in $\mathcal{C}^{(n+1)}$, that is for every $\alpha \subset \{1,\ldots,n\}$ there is a groupoid $\pi(\alpha)$ and morphisms $f_{\alpha\beta}: \pi(\beta) \to \pi(\alpha)$ for $\beta \subset \alpha$ so that all diagrams commute. The main example is if K is an (n+1)-ad and $\pi = \pi_1(K)$ where $\pi_1(K)(\alpha) = \pi_1(K(\alpha))$ and $f_{\alpha\beta} = i_\#: \pi_1(K(\beta)) \to \pi_1(K(\alpha))$.

If π is a groupoid of type 2^n, then there is an (n+1)-ad $K(\pi,1)$ so that $\pi_1(K(\pi,1)) = \pi$. This is defined as follows: the components of $\pi(\alpha)$ are groups G_i, so let $K(\pi,1)(\alpha) = \bigcup_i K(G_i,1)$ and make the corresponding maps inclusions.

We can now state our main theorem:

Theorem 3. Let π be a groupoid of type 2^{n-2}, R a principal ring and $A \subset Wh(\pi\{1,\ldots,n-2\};R)$ a self-dual subgroup. Then there are surgery obstruction groups $L_m^A(\pi;R)$ so that if $\phi: M \to X$ is a normal map between a manifold n-ad M of dimension m and a Poincare n-ad X over R with torsion in A, $\partial_{n-1}\phi$ a homology equivalence over R with torsion in A, $\pi_1(\delta_{n-1}X) = \pi$, $m - n \geq 3$, then there is an obstruction $\sigma(\phi) \in L_m^A(\pi;R)$ so that $\sigma(\phi) = 0$ if and only if ϕ is normally cobordant rel $\partial_{n-1}M$ to a homology equivalence of n-ads over R with torsion in A.

Proof: Assume n = 2. Define $L_m^A(\pi;R) = \bigoplus_i L_m^A(K_i;R)$ where the K_i are the components of a space of type $K(\pi,1)$. If $R = \mathbb{Z}_p$, the result follows from Theorems 1 and 2. For R arbitrary, the result will follow from Lemma 2 provided X is a Poincare n-ad over $\mathbb{Z}_p$.

Let $[X] \in H_m(X;\mathbb{Z})$ be the fundamental class, and C_* the mapping cone chain complex of $[X]\cap$. Since X is a Poincare complex over R with torsion in A, $C_* \otimes R$ is acyclic and so $C_* \otimes \mathbb{Z}_p$ is acyclic. Thus X is a Poincare complex over $\mathbb{Z}_p$ and is easily seen to have torsion in A'. Thus a surgery problem over (π,R,A) is equivalent to a surgery problem over $(\pi, \mathbb{Z}p, A')$. This gives the result.

The n-ad case is similar. Note the orientation homomorphism w was suppressed from the notation.

The theorem works for any ring R that satisfies Lemma 2.

Torsion for arbitrary rings can be bizarre. For example, let $R = \mathbb{R}[x,y]/(x^2+y^2-1)$. Then $Wh(1;R) \cong \mathbb{Z}/2\mathbb{Z} \oplus \mathbb{R}^{\cdot}/\mathbb{Q}^{\cdot}$, but $Wh(1;\mathbb{Z}_P) = 0$ for any P ([D20]).

5.3. Classifying Spaces for Surgery.

In this section we define classifying spaces for surgery groups, as was first done by Quinn [H8], and use these spaces to painlessly derive some properties of surgery groups, notably the long exact sequence of surgery.

Let $\mathcal{C}$ be a small cobordism category (Stong [A16]) and define a Δ-set by

$\hat{\Omega}_*^{\mathcal{C}}(\Delta^k)$ = the set of (n+2)-ads in $\mathcal{C}$, with face maps induced by face maps of objects.

If $\mathcal{C}$ is graded (e.g. manifolds), then define

$\hat{\Omega}_n(\Delta^k)$ = those elements in $\hat{\Omega}_*^{\mathcal{C}}(\Delta^k)$ of dimension $k + n$.

Define $\Omega_n^{\mathcal{C}} = Ex^{\infty}(\hat{\Omega}_n^{\mathcal{C}})$.

According to Prop. 1.4.4 of [H8], $\pi_r(\Omega_n^{\mathcal{C}}) \cong \Omega_{n+r}(\mathcal{C})$, the (n+r)-th cobordism group of the category $\mathcal{C}$. Also $\Omega_n^{\mathcal{C}}$ is an infinite loop space, $\Omega_n^{\mathcal{C}} \simeq \Omega\Omega_{n-1}^{\mathcal{C}}$.

It follows that given $\Pi = (\bar{K},R,A)$ as in Section 5.2, there exist classifying spaces Ω_j^{Π} and $\mathcal{R}_j^{\Pi}$ so that $\pi_m(\Omega_j^{\Pi}) \cong \Omega_{m+j}(\Pi)$ and $\pi_m(\mathcal{R}_j^{\Pi}) = \text{Image}(\mathcal{R}_{m+j}(\Pi) \to \Omega_{m+j}(\Pi))$. There is a natural map

$$\mathcal{R}_j^{\Pi} \to \Omega_j^{\Pi}.$$

Let $\mathbb{L}_j(\Pi)$ deonte the fiber of the map $\mathcal{R}_{j-1}^{\Pi} \to \Omega_{j-1}^{\Pi}$.

Theorem 1. $\mathbb{L}_j(\Pi)$ is an infinite loop space with

$$\pi_m(\mathbb{L}_j(\Pi)) \cong L_{m+j}(\Pi).$$

Define $\partial_i\Pi = (\partial_i\bar{K},R,A)$ where $\partial_i\bar{K} = (\partial_i K,w|\pi_1(|\partial_i K|))$ and similarly for $\delta_i\bar{K}$. Then there are natural maps $\mathbb{L}_j(\partial_i\Pi) \to \mathbb{L}_j(\delta_i\Pi) \to \mathbb{L}_j(\Pi)$, which is, up to homotopy, a fibration. Thus by the long exact homotopy sequence of a fibration, we have

Theorem 2. There is a long exact sequence

$$\cdots \to L_m^A(\partial_i\bar{K};R) \to L_m^A(\delta_i\bar{K};R) \to L_m^A(\bar{K};R) \to L_{m-1}^A(\partial_i\bar{K};R) \to \cdots .$$

These ideas are more fully expounded in Quinn's thesis, [H8] and in an article in the Georgia Conference on the Topology of Manifolds. See also section 17 in Wall [H19].

5.4. The Periodicity Theorem, Part I.

Let N^n be a closed orientable manifold and define $xN:L_m(\Pi) \to L_{m+n}(\Pi)$ by sending $M \to X$ to $MxN \to XxN$. It is easy to check that this is a well-defined homomorphism. For $R = \mathbb{Z}$, this map is determined partially in Wall [H19], Williamson [H20] and Shaneson [H10]. In this section we show that $x\mathbb{C}P^2$ is an isomorphism for the non-simple case, $A = Wh(\pi;R)$. The general case will follow in Section 6.3. Recall that $L_m^A(\mathbb{Z}_P\pi) = L_{m+4}^A(\mathbb{Z}_P\pi)$.

Theorem 1. For $m \geq 5$, $x\mathbb{C}P^2:L_m^h(\pi,w;R) \to L_{m+4}^h(\pi,w;R)$ is an isomorphism, coinciding with the isomorphism above if $R = \mathbb{Z}_P$.

Proof: Even dimensional case, $m = 2k$.

Let $\phi:M \to X$ represent an element $x \in L_m^h(\pi,w;R)$. Assume as in Section 5.2 that $R = \mathbb{Z}_P$. Using $R\pi$-coeffients throughout, we can assume ϕ is k-connected and $K_k(M)$ is free. Then, algebraically, x is represented by a Hermitian form on $K_k(M)$. Let $f_i:S^k xD^k \to M$, $i = 1,\ldots,r$, be immersions representing a basis for $K_k(M)$.

Multiplying by $\mathbb{C}P^2$, the only non-vanishing kernel groups are $K_k(Mx\mathbb{C}P^2)$, $K_{k+2}(Mx\mathbb{C}P^2)$ and $K_{k+4}(Mx\mathbb{C}P^2)$, all isomorphic to $K_k(M)$.

Let $j:S^2 \to \mathbb{C}P^2$ be an embedding representing a generator of $\pi_2(\mathbb{C}P^2) \cong \mathbb{Z}$. Define $g_i:S^k xS^2 \to Mx\mathbb{C}P^2$ by

$$S^k \times S^2 \xrightarrow[(x,y)\mapsto(x,1,y)]{} S^k \times D^k \times S^2 \xrightarrow{(f_i,j)} M\times\mathbb{C}P^2,$$ and assume they are in general position.

It follows easily from Theorem 1.1.6 and Spanier [A15], Chapter 5, that $\lambda(f_i,f_{i'}) = \lambda(g_i,g_{i'})$, $\mu(f_i) = \mu(g_i)$ since j is an embedding representing a generator. The spheres $g_i(S^k\times 1)$ are disjointly embedded and framed, so we can do surgery on them, obtaining a manifold N. Let W be the trace of the surgery and $\psi: N \to X$. Then

$$K_i(W, M\times\mathbb{C}P^2) = \begin{cases} K_k(M\times\mathbb{C}P^2) & i = k+1 \\ 0 & \text{otherwise} \end{cases}$$

and so $K_i(W) = 0$ for $i \neq k+2,\ k+4$.

We have

$$\begin{array}{ccccc} K_{k+4}(M\times\mathbb{C}P^2) & \cong & K_{k+4}(W) & \to & K_{k+4}(W,N) \\ & \searrow\cong & & & \downarrow\cong \\ & & K_{k+1}(W,M\times\mathbb{C}P^2) & \overset{\cong}{\to} & K^{k+1}(W,M\times\mathbb{C}P^2) \end{array}$$

so $K_{k+4}(N) = 0$, and the only non-vanishing kernel is $K_{k+2}(N) \cong K_k(M)$.

Surgery on the spheres $S^k\times 1$ yielded immersions $h_i: S^{k+2} \to M\times CP^2$; furthermore $\lambda(h_i,h_j) = \lambda(f_i,f_j)$, $\mu(h_i) = \mu(f_i)$ since the spheres $g_i(S^k\times 1)$ are disjointly embedded. Clearly the maps h_i represent a basis for $K_{k+2}(N)$ and correspond to the f_i under the isomorphism $K_{k+2}(N) \cong K_k(M)$. Also, $K_{k+2}(N) \cong \pi_{k+3}(\psi)$ and so the h_i are framed. Thus the surgery obstruction for $M\times\mathbb{C}P^2 \to X\times\mathbb{C}P^2$ is represented by the Hermitian form on $K_k(M)$.

<u>Odd-dimensional case, $m = 2k-1$</u>: Let $\phi: M \to X$ represent $x \in L_m^h(\pi, w; R)$, $R = \mathbf{Z}_p$, and as in Section 4.3, x is represented algebraically by the subkernels $K_k(U, \partial U)$ and $K_k(M_0, \partial U)$ in $K_{k-1}(\partial U)$, where

$$U = \bigcup_{i=1}^{r} f_i(S^{k-1} \times D^k),\ f_i \text{ disjoint embeddings re-}$$

resenting generators of $K_{k-1}(M)$, $M_0 = M - \mathrm{Int}(U)$.

Do surgery on the embedded spheres $f_i(S^{k-1} \times 1) \times pt.$ in $M \times \mathbb{C}P^2$ to get a manifold N. Then $K_{k-1}(N) = 0$, and as above, the surgery yields framed embeddings $g_i: S^{k+1} \to N$ from $S^{k-1} \times S^2 \to M \times CP^2$. The maps g_i generate $K_{k+1}(N) \cong K_{k+1}(M \times \mathbb{C}P^2) \cong K_{k-1}(M)$. Let W be the trace of the surgery.

Then $K_k(W) \cong K_k(M \times \mathbb{C}P^2, U) \cong K_k(M, U)$; also $K_k(N) \cong K_k(W)$. But $K_k(M, U) \cong K_k(M_0, \partial U)$, so $K_k(N)$ is free. Do surgery on framed spheres representing a basis and assume these spheres are disjoint from the $g_i(S^{k+1})$; let Q be the resultant. Clearly

$$K_i(Q) = \begin{cases} K_{i-2}(M) & i = k+1,\ k+3 \\ 0 & \text{otherwise.} \end{cases}$$

The embeddings g_i determine embeddings (also denoted g_i) in Q; these maps generate $K_{k+1}(Q) \cong K_{k+1}(N)$. Let $V = \bigcup_{i=1}^{r} g_i(S^{k+1} \times D^{k+2}) \subset Q$. Then the map $K_{k-1}(\partial U) \to K_{k+1}(\partial V)$, $f_i(S^{k-1} \times 1) \mapsto g_i(S^{k+1} \times 1)$, $f_i(1 \times S^{k-1}) \mapsto g_i(1 \times S^{k-1})$, is an isomorphism of kernels. This isomorphism sends $K_k(U, \partial U)$ to $K_{k+2}(V, \partial V)$; we must show it sends $K_k(M_0, \partial U)$ to $K_{k+2}(Q_0, \partial V)$, $Q_0 = Q - \mathrm{Int}(V)$.

To this end, note we have

$$\begin{array}{ccccc}
K_{k-1}(\partial U) & \overset{\cong}{\to} & K_{k+1}(\partial U \times P) & \overset{\cong}{\to} & K_{k+1}(\partial(U \times T)) \\
\uparrow & & \uparrow & & \nwarrow \\
K_k(M_0,\partial U) & \to & K_{k+2}(M_0 \times \mathbb{C}P^2, \partial U \times \mathbb{C}P^2) & & \\
 & & \downarrow \cong & & \\
 & & K_{k+2}(M \times \mathbb{C}P^2, U \times \mathbb{C}P^2) & & \\
 & & \uparrow \cong & & \\
 & & K_{k+2}(M \times \mathbb{C}P^2, U \times D) & \longrightarrow & K_{k+2}(M \times \mathbb{C}P^2 - U \times D, \partial(U \times D))
\end{array}$$

where D is a regular neighborhood of S^2 in $\mathbb{C}P^2$. Doing surgery on the f_i inside of UxD instead of $U \times \mathbb{C}P^2$ we get a manifold V', each component of which has the homotopy type of $S^2 \vee S^{k+1} \vee S^{k+3}$ (the S^2 comes from D, the S^{k+1} is constructed as before and the S^{k+3} is the transverse sphere). We can assume $V \subset \operatorname{Int}(V')$. The inclusion of ∂V and $\partial V'$ in $V' - \operatorname{Int}(V)$ induces isomorphisms on $K_{k+1}(\ )$ as is easily seen by the Mayer-Vietoris sequence of the triple $(V';V,V'-\operatorname{Int}(V);\partial V)$.

Now $K_{k+1}(\partial(U \times D)) \cong K_{k+1}(\partial V')$ and the map $K_{k+1}(\partial U \times \mathbb{C}P^2) \to K_{k+1}(\partial(U \times D))$ is an isomorphism. (It is onto by the fact $K_{k+1}(U \times \partial D) = 0$ since $\partial D \to S^2$ is the non-trivial S^1-bundle, $\partial D = S^3$ and the homology sequence. Since both are free of the same rank, it is an isomorphism.) Now we identify

$$\begin{aligned}
K_{k+2}(M \times \mathbb{C}P^2 - \operatorname{Int}(U \times D), \partial(U \times D)) &\cong K_{k+2}(N - \operatorname{Int}(V'), \partial V') \\
&\cong K_{k+2}(N - \operatorname{Int}(V), \partial V) \\
&\cong K_{k+2}(Q_0, \partial V).
\end{aligned}$$

This concludes the proof.

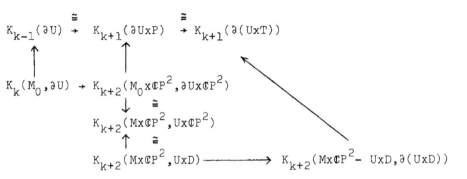

Theorem 2. *If K is an n-ad and $m - n > 3$, then $\times \mathbb{C}P^2$ is an isomorphism* $L^h_m(K,w;R) \to L^h_{m+4}(K,w;R)$.

Proof: Immediate by induction, the five lemma, Theorem 5.3.2 and Theorem 1.

Chapter 6. Relations Between Surgery Theories.

6.1. The Long Exact Sequence of Surgery with Coefficients.

Let π be a multiplicative group, and $w:\pi \to \{\pm 1\}$ a homomorphism. For convenience, let $\bar{\pi} = (\pi, w)$. Suppose A is a self-dual subgroup of $Wh(\pi;R)$.

Def. Let $CH_n^A(\bar{\pi};R)$ denote the subgroup of $L_n^A(\bar{\pi};R)$ of obstructions realizable by normal maps $M \to X$ with $\partial M = \emptyset = \partial X$, and M and X H-manifolds (H = TOP, PL, or DIFF).

For example, $CPL_n(1,\mathbb{Z}) = L_n(1;\mathbb{Z})$ by Theorem 4.4.2.

Theorem 1. There is an exact sequence

$$0 \to CH_{n+1}^A(\bar{\pi};R) \to L_{n+1}^A(\bar{\pi};R) \to \mathcal{R}_n^H(\bar{\pi},A,R) \to \\ \to \Omega_n^H(\bar{\pi},A,R) \to L_n^A(\bar{\pi};R) \to 0.$$

Proof: The exactness of this sequence at the last four terms follows from Corollary 5.2.1. Let β denote the map $\mathcal{R}_n^H(\bar{\pi},A,R) \to \Omega_n^H(\bar{\pi},A,R)$. We show that $\ker(\beta) \cong L_{n+1}^A(\bar{\pi};R)/CH_{n+1}^A(\bar{\pi};R)$.

Let $x \in \ker(\beta)$. Then x is represented by a homology equivalence over R, $M \to X$, bounded by a normal map $N \to Y$. This gives a well-defined map $\ker(\beta) \to L_{n+1}^A(\bar{\pi};R)/\sim$, where $\sim$ is the equivalence relation defined by $\theta_1 \sim \theta_2$ if θ_i is represented by $N_i \to Y_i$, $i = 1,2$, so that $\partial N_1 \to \partial Y_1$

and $\partial N_2 \to \partial Y_2$ represent the same class in $\mathcal{R}_n^H(\bar{\pi},A,R)$.

Suppose $\theta_1 \sim \theta_2$ as above. Then if $W \to V$ is a homology equivalence over R which bounds $\partial N_1 \to \partial Y_1$ and $\partial N_2 \to \partial Y_2$, define $N = N_1 \cup W \cup (-N_2)$, $Y = Y_1 \cup V \cup (-Y_2)$ and extend the respective maps to get $N \to Y$. Since $W \to V$ is a homology equivalence over R, the surgery obstruction of $N \to Y$ is $\theta_1 - \theta_2$. Thus $\theta_1 - \theta_2 \in CH_{n+1}^A(\bar{\pi};R)$. Conversely, $\theta_1 - \theta_2 \in CH_{n+1}^A(\bar{\pi};R)$ implies $\theta_1 \sim \theta_2$.

So we have a well-defined map $\ker(\beta) \to L_{n+1}^A(\bar{\pi};R)$ modulo $CH_{n+1}^A(\bar{\pi};R)$. Straightforward calculations show that this is a homomorphism and in fact an isomorphism.

This sequence is sort of a universal version of the sequence in Theorem 4.5.1. A similar sequence can be derived for n-ads using Theorem 5.3.2.

Let $g:R \to R'$ be a ring homomorphism, where R and R' are principal rings; let π and π' be multiplicative groups with homomorphisms w,w' as usual, and let $f:\pi \to \pi'$ be a homomorphism so that $w'f = w$. Then there is a $\mathbb{Z}/2\mathbb{Z}$-equivariant map $g[f]_*:Wh(\pi;R) \to Wh(\pi;R')$ induced by $g[f]:R\pi \to R'\pi'$, $g[f](rx) = g(r)f(x)$. We want to study the effect on surgery groups of the map $g[1]:R\pi \to R'\pi$.

Let A, A' be self-dual subgroups of $Wh(\pi;R)$, $Wh(\pi;R')$ so that $g[1]_*A \subset A'$. Let $\Pi = (\pi,w,R,A)$, $\Pi' = (\pi,w,R',A')$.

Using the notation of Section 5.2, let $\mathcal{R}_m(\Pi',\Pi) = h_{\Pi'}^{-1}\partial^{-1}h_\Pi Q_{m-1}(\Pi)$.

Define $\mathcal{L}_m(g;\Pi,\Pi') = \mathcal{R}_m(\Pi',\Pi)/g_*\mathcal{R}_m(\Pi)$, where $g_*:Q_m(\Pi) \to Q_m(\Pi')$ is the induced map. By the remarks following Corollary 5.2.1, we can assume all spaces involved to be manifolds.

Theorem 2. There is a long exact sequence

$$\cdots \longrightarrow \mathcal{L}_m(g;\Pi,\Pi') \xrightarrow{j_*} L_m^A(\pi,w;R) \xrightarrow{L(g)} L_m^{A'}(\pi,w;R') \xrightarrow{\partial_*} \mathcal{L}_{m-1}(g;\Pi,\Pi') \to \cdots.$$

Proof: This can be proved using classifying spaces and showing that the fiber of $\mathbb{L}_m(\pi,w,R,A) \to \mathbb{L}_m(\pi,w,R',A')$ has the required properties; we give here an elementary geometric proof.

Define $L(g)$ to be the functorial map; ∂_* is defined by sending $f:M \to X$ in $L_m^{A'}(\pi,w;R')$ to the class of $\partial f:\partial M \to \partial X$, a homology equivalence over R'. We define $j_*[f:M \to X]$ to be the class of f in $L_m^A(\pi,w;R)$. Elementary considerations show that these are well-defined. Recall that a homology equivalence over a ring is always assumed to be normal. We omit mention of A, A', and w throughout.

(i). $L(g)j_* = 0$: Clearly if $f:M \to X$ is a homology equivalence over R', then f represents 0 in $L_m(\pi;R')$.

(ii). $\partial_*L(g) = 0$: $\partial_*L(g)[f:M \to X]$ is represented by $\partial f:\partial M \to \partial X$, a homology equivalence over R, and thus is 0.

(iii). $j_*\partial_* = 0$: $j_*\partial_*[f:M \to X]$ is represented by $\partial f:\partial M \to \partial X$ in $L_m(\pi;R)$, and f gives a bordism of ∂f to a homology equivalence over R.

(iv). $\ker(L(g)) \subset \mathrm{Im}(j_*)$: Suppose $f:M \to X$ is a normal map, f a homology equivalence over R, and f is cobordant to a homology equivalence over R', i.e. $[f] \in \ker(L(g))$, $f':M' \to X'$. Then $j_*[f'] = [f]$.

(v). $\ker(\partial_*) \subset \mathrm{Im}(L(g))$: If $\partial_*[f:M \to X] = 0$, then ∂f is cobordant to a homology equivalence over R, the cobordism given by a homology equivalence over R', $F:N \to Y$, and the cobordism extension property (pg. 45 in [J22]) now shows that $L(g)[f \cup F : M \cup_{\partial N} N \to X \cup_{\partial X} Y] = [f:M \to X]$.

(vi). $\ker(j_*) \subset \mathrm{Im}(\partial_*)$: Let $j_*[f:M \to X] = 0$. Then f is cobordant to a homology equivalence over R by a cobordism $F:N \to Y$. Then $\partial_*[F] = [f]$. This completes the proof.

Remarks: (1) A sequence of this type was first used by Pardon [K9] to relate $L^h_m(\pi)$ and $L^h_m(\mathbb{Q}\pi)$, where he defines $L^h_m(\mathbb{Q}\pi)$ to be the usual cobordism group, except boundaries are allowed to change by rational h-cobordisms. The correction factor $\mathcal{L}_m$ in this case is shown to be $\mathcal{R}_m(\Pi',\Pi)$, where $\Pi = (\pi,w,\mathbb{Z},Wh(\pi))$, $\Pi' = (\pi,w,\mathbb{Q},Wh(\pi,\mathbb{Q}))$.

(2) This is also related to the surgery of Cappell and Shaneson [K5]. In fact, if $\mathbb{Z}\pi \to R\pi$ and $g[1]:R\pi \to R'\pi$ are locally epic, then $\mathcal{L}_m(g;\Pi,\Pi') \cong \Gamma^h_{m+1}(\phi)$, where ϕ is the diagram

$$\begin{array}{ccc} \mathbb{Z}\pi & \longrightarrow & R\pi \\ = \big\downarrow & & \big\downarrow \\ \mathbb{Z}\pi & \longrightarrow & R'\pi \end{array}$$

and $\Pi = (\pi,w,R,Wh(\pi,R))$ etc.

(3) A similar formulation can be done for n-ads.

6.2. The Rothenberg Sequence.

In this section we generalize the sequence of Rothenberg (unpublished) and Shaneson [H9] to arbitrary coefficients using the exact sequence of Section 6.1. We begin with a series of lemmas. The first of these gives the proof of Theorem 2.3 in the light of Chapter 4.

Lemma 1 (Theorem 2.3) Let $x \in Wh(\pi;\mathbb{Z}_p)$ and M^n a manifold with $\pi_1(M) = \pi$, $n \geq 5$. Then there is an h-cobordism $(W;M,M')$ over $\mathbb{Z}_p$ with $\tau(W,M;\mathbb{Z}_p) = x$. Furthermore there is a map $q:W \to M$ so that $qi = 1$, where $i:M \subset W$; q is a homology equivalence over $\mathbb{Z}_p$.

Proof: Let A be a kxk matrix representing x. Add k trivial 2-handles to MxI to get a manifold triad $(W_+;M,M_+)$ and a normal map $\phi_+:(W_+;M,M_+) \to (MxI;Mx0,Mx1)$, since this amounts to surgery on the identity map. Then $\pi_2(W_+,M_+)$ is a free $\mathbb{Z}\pi$-module on generators $e_1,\ldots,e_k$. Also, $\pi_2(W_+,M_+) = \pi_2(\phi_+|(W_+,M_+))$, so every element in $\pi_2(W_+,M_+)$ is represented by a framed embedding $S^2 \subset M_+$, since $\dim M_+ = n \geq 5$.

Let $A = (a_{ij})$ and assume $a_{ij} \in \mathbb{Z}\pi$ (if not, multiply A by a suitable factor). Do surgery on spheres representing

$$\sum_{j=1}^{k} a_{ij}e_j,\quad i = 1,\ldots,k, \text{ and let } \phi:(W;M,M') \to (MxI;Mx0,Mx1)$$

be the trace of the surgeries. Then ϕ is a homology equivalence over $\mathbb{Z}_p$, and so is an h-cobordism over $\mathbb{Z}_p$, $\tau(W,M;\mathbb{Z}_p) = x$, and $W \to MxI \to M$ is the desired map q.

Lemma 2. *Let* $f: M^n \to X^n$ *be a homology equivalence over* $\mathbb{Z}_p$ *between manifolds*, $n \geq 5$; *let* $a \in Wh(\pi_1(X);\mathbb{Z}_p)$. *Then* f *is cobordant, by a homology equivalence over* $\mathbb{Z}_p$, *to a homology equivalence over* $\mathbb{Z}_p$ *with torsion* a *if and only if* $\tau(f;\mathbb{Z}_p) = b + (-1)^{n+1}b^* + a$ *for some* $b \in Wh(\pi_1(X);\mathbb{Z}_p)$.

Proof: Let $\pi = \pi_1(X)$. We identify all relevant Whitehead groups by their respective maps (e.g. $Wh(\pi_1(M);\mathbb{Z}_p)$ and $Wh(\pi_1(X);\mathbb{Z}_p)$ are identified by f_*).

Suppose $\tau(f;\mathbb{Z}_p) = b + (-1)^{n+1}b^* + a$. Let $(W;M,M')$ be an h-cobordism over $\mathbb{Z}_p$ with $\tau(W,M;\mathbb{Z}_p) = b$ and $q: W \to M$ the map constructed in Lemma 1. Define $F: W \to X \times I$ by $F(w) = (fq(w), \phi(w))$, where ϕ is a Urysohn function. Then the diagram

$$\begin{array}{ccc} M & \xrightarrow{\quad f \quad} & X\times 0 \\ \cap & & \cap \\ W & \xrightarrow{\quad F \quad} & X\times I \\ \cup & & \cup \\ M' & \xrightarrow{\quad f' \quad} & X\times 1 \end{array} \qquad (*)$$

commutes, so

$$\begin{aligned} \tau(f';\mathbb{Z}_p) &= \tau(F;\mathbb{Z}_p) + \tau(W,M';\mathbb{Z}_p) \\ &= \tau(f;\mathbb{Z}_p) - \tau(W,M;\mathbb{Z}_p) + \tau(W,M';\mathbb{Z}_p) \\ &= b + (-1)^{n+1}b^* + a - b + (-1)^n b^* \qquad \text{by Theorem 3.1.4} \\ &= a. \end{aligned}$$

Now suppose we have a diagram (*) with f', F homology equivalences over $\mathbb{Z}_p$, $\tau(f';\mathbb{Z}_p) = a$. Then

$$\begin{aligned} \tau(f;\mathbb{Z}_p) &= \tau(F;\mathbb{Z}_p) + \tau(W,M;\mathbb{Z}_p) \\ &= \tau(f';\mathbb{Z}_p) - \tau(W,M';\mathbb{Z}_p) + \tau(W,M;\mathbb{Z}_p) \end{aligned}$$

$$= a + \tau(W,M;\mathbb{Z}_p) + (-1)^{n+1}\tau(W,M;\mathbb{Z}_p)^*.$$

Lemma 3. Let $a \in Wh(\pi;\mathbb{Z}_p)$, $a = (-1)^{n+1}a^*$. Then there is a homology equivalence over $\mathbb{Z}_p$, $f:M^n \to X^n$, M and X manifolds, with $\tau(f;\mathbb{Z}_p) = a$. Conversely, if $f:M^n \to X^n$ is a homology equivalence over $\mathbb{Z}_p$, M and X manifolds, $n \geq 5$, then $\tau(f;\mathbb{Z}_p) = (-1)^{n+1}\tau(f;\mathbb{Z}_p)^*$.

Proof: Let a be represented by a matrix A and define normal maps $f':M' \to X'$, $F:W \to X' \times I$ as follows:

n odd: The matrix

$$\begin{bmatrix} A & 0 \\ 0 & (A^*)^{-1} \end{bmatrix}$$

represents the surgery obstruction of some normal $f_0:M_0 \to X'$; f_0 is cobordant to a homology equivalence over $\mathbb{Z}_p$ since the above matrix represents 0 in $L_n^h(\pi,w;\mathbb{Z}_p)$. Let B be a matrix representing $\tau(f';\mathbb{Z}_p)$, where f' is a homology equivalence over $\mathbb{Z}_p$ cobordant to f_0. Then we can construct a $(-1)^{(n+1)/2}$-Hermitian form (G,λ,μ) with

$$A_\lambda = \left[\begin{array}{cc|cc} A & 0 & \multicolumn{2}{c}{\multirow{2}{*}{0}} \\ 0 & (A^*)^{-1} & & \\ \hline \multicolumn{2}{c|}{\multirow{2}{*}{0}} & (B^*)^{-1} & 0 \\ & & 0 & B \end{array}\right].$$

Let F be a normal map with surgery obstruction (G,λ,μ).

n even: Apply A to the basis of the standard kernel to get

a $(-1)^{n/2}$-Hermitian form (G,λ,μ); let $f_0 : M_0 \to X'$ realize this form. Then f_0 is cobordant to a homology equivalence $f' : M' \to X'$ with torsion represented by B. Let F represent

$$\begin{bmatrix} A & 0 \\ 0 & (B^*)^{-1} \end{bmatrix} \text{ in } L^h_{n+1}(\pi,w;\mathbb{Z}_p).$$

Then $f' + \partial F : M' + \partial W \to X' + \partial(X' \times I)$ has torsion a.

To show the converse, suppose $\tau(f;\mathbb{Z}_p) = a$. Construct h-cobordisms over $\mathbb{Z}_p$, $(W;M,M')$ and $(Y;X,X')$, with $\tau(W,M;\mathbb{Z}_p) = a$, $\tau(Y,X;\mathbb{Z}_p) = -a$. Then the space $N = W \bigcup_f Y$ is a simple Poincare complex over $\mathbb{Z}_p$. This follows from the construction of W and Y. Using this we get

$$\begin{aligned} (-1)^n a^* &= \tau(W,M';\mathbb{Z}_p) \\ &= \tau(N,M';\mathbb{Z}_p) \\ &= (-1)^n \tau(N,X';\mathbb{Z}_p)^* \\ &= (-1)^n \tau(Y,X';\mathbb{Z}_p)^* \\ &= (-1)^n ((-1)^{n+1} a^*)^* \\ &= -a \end{aligned}$$

by repeated use of Theorem 3.1.4.

We can now prove our main theorem.

Theorem 1. Let $A \subset B \subset Wh(\pi;\mathbb{Z}_p)$ be self-dual subgroups where * is defined by $w : \pi \to \{\pm 1\}$. Then there is a long exact sequence

$$\cdots \to H^{n+1}(\mathbb{Z}/2\mathbb{Z};B/A) \to L^A_n(\pi,w;\mathbb{Z}_p) \to L^B_n(\pi,w;\mathbb{Z}_p) \to H^n(\mathbb{Z}/2\mathbb{Z};B/A) \to \cdots$$

terminating at $H^5(\mathbb{Z}/2\mathbb{Z};B/A)$.

Here $H^*(\mathbb{Z}/2\mathbb{Z};B/A)$ denotes cohomology of groups, [A26].

Proof: We show that $\mathcal{L}_n(1;\Pi,\Pi') = H^{n+1}(\mathbb{Z}/2\mathbb{Z};B/A)$ where $\Pi = (\pi,w,\mathbb{Z}_p,A)$, $\Pi' = (\pi,w,\mathbb{Z}_p,B)$. The result will then follow from Theorem 6.1.2.

Recall that $H^{n+1}(\mathbb{Z}/2\mathbb{Z};B/A) = \dfrac{\{b \in B:\ b = (-1)^{n+1}b^*\}}{\{b+(-1)^{n+1}b^*: b\in B\} + A}$, and define

$$\Phi: \mathcal{L}_n \to H^{n+1}(\mathbb{Z}/2\mathbb{Z};B/A)$$

by $[f:M \to X] = \tau(f;\mathbb{Z}_p)$.

Φ is well-defined and injective by Lemma 2. (Note we need only consider cobordisms $F:W \to X{\times}I$ by the remarks following Corollary 5.2.1, Theorem 6.1.2 and the five lemma.) Φ is surjective by Lemma 3.

We interpret the maps in the sequence as follows:

$L^B_{2k} \to H^{2k}$ sends $[G,\lambda,\mu]$ to $[A_\lambda]$;

$H^{2k} \to L^A_{2k-1}$ sends $[Q]$ to the class of $\begin{bmatrix} Q & 0 \\ 0 & (Q^*)^{-1} \end{bmatrix}$;

$L^B_{2k-1} \to H^{2k-1}$ sends $[\alpha]$ to $[\tau(\alpha;\mathbb{Z}_p)]$;

$H^{2k-1} \to L^A_{2k-2}$ sends $[Q]$ to the $(-1)^{k-1}$-Hermitian form defined by applying Q to the basis of the standard kernel.

6.3. The Periodicity Theorem, Part II.

In this section we use the Rothenberg sequence to complete the periodicity theorem.

Theorem 1. For $m \geq 5$, $\times \mathbb{C}P^2 : L_m^A(\pi,w;R) \to L_{m+4}^A(\pi,w;R)$ is an isomorphism.

Proof: By Theorem 6.2.1 and the five lemma we need only show that $\times \mathbb{C}P^2$ induces an isomorphism $H^m(\mathbb{Z}/2\mathbb{Z};B/A) \to H^{m+4}(\mathbb{Z}/2\mathbb{Z};B/A)$, where we interpret these groups geometrically as in Section 6.2.

Let $f:M \to X$ be a homology equivalence over R and C_* the chain complex of f as in Chapter 2. Since $\mathbb{C}P^2$ has a cell decomposition $e^0 \cup e^2 \cup e^4$, the chain complex of $f \times 1_{\mathbb{C}P^2}$ is given by $C_* + C_{*+2} + C_{*+4}$. Thus $\tau(f\times 1;\mathbb{Z}_p) = 3\tau(f;\mathbb{Z}_p)$. But $H^m(\mathbb{Z}/2\mathbb{Z};B/A)$ is 2-torsion, so this is an isomorphism.

Corollary 1. For $m > 5$, $\times \mathbb{C}P^2 : \mathcal{L}_m(g;\Pi,\Pi') \to \mathcal{L}_{m+4}(g;\Pi,\Pi')$ is an isomorphism.

6.4. Simple Linking Numbers.

Let π be a finite group and $w:\pi \to \{\pm 1\}$ a homomorphism. Let $K(\pi)$ denote the kernel of the map $Wh(\pi) \to Wh(\pi,\mathbb{Q})$. In this section we will define a K-theoretic group of simple linking forms which determines the kernel of $L_n^{K(\pi)}(\pi,w) \to L_n^s(\pi,w;\mathbb{Q})$ for n odd. This also estimates the

the kernel of the corresponding map with $K(\pi)$ replaced by s.

Let G be a free and based $\mathbb{Z}\pi$-module and λ a $(-1)^k$-Hermitian form on G which is non-degenerate, i.e. $A_\lambda : G \to G^*$ is injective. We further assume that $A_\lambda \otimes 1 : G \otimes \mathbb{Q} \to G^* \otimes \mathbb{Q}$ is a simple isomorphism. Let Λ denote $\mathbb{Q}\pi/\mathbb{Z}\pi$; let $I_k = \langle x+(-1)^k x^* : x \varepsilon \mathbb{Z}\pi \rangle$.

Lemma 1. We can associate to (G,λ) a triple (K,λ_0,μ_0) where

(i) K is a torsion group,

(ii) $\lambda_0 : K x K \to \Lambda$ is $(-1)^k$-Hermitian and non-singular,

(iii) $\mu_0 : K \to \mathbb{Q}\pi/I_k$ is an associated quadratic form.

Proof; Let $K = \operatorname{coker}(A_\lambda)$; then there exists r so that $rG^* \subset \operatorname{Im} A_\lambda$. Define $\lambda_0' : G^* x G^* \to \mathbb{Q}\pi$ by

$$\lambda_0'(x,y) = \frac{1}{r^2}\lambda(A_\lambda^{-1} rx, A_\lambda^{-1} ry).$$

For $x,y \ \varepsilon \ \operatorname{Im} A_\lambda$,

$$\lambda_0'(x,y) = \lambda(A_\lambda^{-1}x,\ A_\lambda^{-1}y)$$
$$= x(A_\lambda^{-1}y) \ \varepsilon \ \mathbb{Z}\pi,$$

so the map $\lambda_0 : K x K \to \Lambda$, $\lambda_0(f(x),f(y)) = \lambda_0'(x,y) \bmod \mathbb{Z}\pi$, $f : G^* \to K$, is well-defined. It is easily verified that λ_0 is $(-1)^k$-Hermitian.

Applying $\operatorname{Hom}_{\mathbb{Z}\pi}(\ \ ,\mathbb{Z}\pi)$ to the sequence

$$0 \to G \to G^* \to K \to 0$$

we get

$$0 = \operatorname{Ext}^1(G^*,\mathbb{Z}\pi) \to \operatorname{Ext}^1(K,\mathbb{Z}\pi) \to G^* \to G^{**} \to \operatorname{Hom}(K,\mathbb{Z}\pi) = 0.$$

Therefore $\operatorname{coker}(A_\lambda^*) \cong \operatorname{Ext}^1(K,\mathbb{Z}\pi)$.

Applying $\operatorname{Hom}_{\mathbb{Z}\pi}(K,\ \ )$ to $0 \to \mathbb{Z}\pi \to \mathbb{Q}\pi \to \Lambda \to 0$ gives

$$0 \to K^* \to \mathrm{Hom}(K,\mathbb{Q}\pi) \to \mathrm{Hom}(K,\Lambda) \to \mathrm{Ext}^1(K,\mathbb{Z}\pi) \to \mathrm{Ext}^1(K,\mathbb{Q}\pi) = 0$$

and so $\mathrm{Hom}(K,\mathbb{Q}\pi) \cong \mathrm{Ext}^1(K,\mathbb{Z}\pi)$.

Since $A_\lambda = A_\lambda^*\Phi$, $\Phi:G \to G^{**}$, $\Phi(g)(x) = x(g)^*$, we have

$K = \mathrm{coker}(A_\lambda)$

$= \mathrm{coker}(A_\lambda^*\Phi)$

$= \mathrm{coker}(A_\lambda^*)$

$\cong \mathrm{Hom}(K,\Lambda)$. This map $K \to \mathrm{Hom}(K,\Lambda)$ is equal to A_{λ_0}, so λ_0 is non-singular.

Define $\mu_0:K \to \mathbb{Q}\pi/I_k$ by $\mu_0(f(x)) = \lambda_0'(x,x) \bmod I_k$. We leave it to the reader to show that μ_0 is well-defined, $\mu_0(xa) = a^*\mu_0(x)a$, and $\lambda_0(x,y) + (-1)^k\lambda_0(x,y)^* = \mu_0(x+y) - \mu_0(x) - \mu_0(y)$.

(K,λ_0,μ_0) is called a standard simple linking form.

If (K,λ_0,μ_0) is a triple, $\lambda_0:K\times K \to \Lambda$ $(-1)^k$-Hermitian, $\mu_0:K \to \mathbb{Q}\pi/I_k$ an associated quadratic form, then we say (K,λ_0,μ_0) is a simple linking form if

(i) $(K,\lambda_0,\mu_0) \oplus (K,-\lambda_0,-\mu_0)$ is standard,

(ii) there is an injective map $f:F_1 \to F_2$, F_1, F_2 free and based so that $f\otimes 1:F_1\otimes\mathbb{Q} \to F_2\otimes\mathbb{Q}$ is a simple isomorphism and $K = \mathrm{coker}(f)$.

Define the group of simple linking forms, $\mathcal{L}^s_{2k-1}(\pi,w)$, to be the monoid of simple linking forms modulo the monoid of standard simple linking forms.

Let $\partial_*:L^s_{2k}(\pi,w;Q) \to \mathcal{L}_{2k-1}(1;\Pi,\Pi')$ denote the boundary homomorphism of Theorem 6.1.2, where $\Pi = (\pi,w,\mathbb{Z},K(\pi))$, $\Pi' = (\pi,w,\mathbb{Q},0)$. Then $\mathrm{coker}(\partial_*)$ is the cobordism group of simple rational homology equivalences modulo cobordism to a homology equivalence over $\mathbb{Z}$ with torsion in $K(\pi)$ (by any normal cobordism). We will show $\mathcal{L}^s_{2k-1}(\pi,w) \cong \mathrm{coker}(\partial_*)$.

Lemma 2. Let $f:M \to X$ be a rational homology equivalence, M a manifold of dimension $2k-1 \geq 5$. Assume f is $(k-1)$-connected. Then "linking" and "self-linking" numbers define a $(-1)^k$-Hermitian form on $K_{k-1}(M)$, with associated quadratic form, over Λ.

Proof: The exact sequence $0 \to \mathbb{Z}\pi \to \mathbb{Q}\pi \to \Lambda \to 0$ yields $0 = K_k(M;\mathbb{Q}\pi) \to K_k(M;\Lambda) \to K_{k-1}(M) \to K_{k-1}(M;\mathbb{Q}\pi) = 0$. Let $K = K_{k-1}(M)$; then $K^* = K^k(M)$. As in Lemma 1, $\mathrm{Hom}(K;\Lambda) = \mathrm{Ext}^1(K;\mathbb{Z}\pi)$, and we get

$$0 \to \mathrm{Hom}(K;\Lambda) \xrightarrow{\cong} K^{k-1}(M;\Lambda) \to K^k(M) = K^* \to 0.$$

This defines a map $K \to \mathrm{Hom}(K,\Lambda)$ and so a non-singular map $\lambda_0:K\mathrm{x}K \to \Lambda$.

Geometrically, we can define this as follows: let $x,y \in K$ be represented by $(k-1)$-spheres S_x, S_y embedded disjointly in M; $rx = 0$ for some r so rS_x bounds a k-chain σ in M. Then $\lambda_0(x,y) = \frac{1}{r}\lambda(S_y,\sigma)$, where λ is the intersection pairing. Similarly for μ_0. It is easily checked that this satisfies the properties alluded to above.

The first of our main technical results is:

Theorem 1. Let $f:M \to X$ be as above. Then f is cobordant to a homology equivalence over $\mathbb{Z}$ with torsion in $K(\pi)$ iff the linking form on $K_{k-1}(M)$ is standard.

Proof: Suppose $F:N \to X \times I$ is a cobordism from f to the $K(\pi)$ homology equivalence over $\mathbb{Z}$, $f':M' \to X$. We may assume F is k-connected and the exact sequence of the pair (N,M) reduces to

$$0 \to K_k(M) \to K_k(N) \to K_k(N,M) \to K_{k-1}(M) \to 0.$$

By adding trivial handles we can assume $K_k(N)$ and $K_k(N,M)$ are free and of the same rank, since $K_i(M)$ is a torsion group, $i = k-1, k$.

Furthermore, $K_k(N,M) \cong K_k(N,\partial N) \cong K^k(N) \cong K_k(N)^*$, and so the middle map above defines a pairing λ on $K_k(N)$. This is defined geometrically in Section 4.3 and A_λ is a simple isomorphism over $\mathbb{Q}\pi$. The form on $K_k(N)/K_k(M)$ induces the linking form on $K_{k-1}(M)$ and so $K_{k-1}(M)$ is standard.

Conversely, assume we have an exact sequence

$$0 \to G \xrightarrow{A_\lambda} G^* \xrightarrow{\phi} K_{k-1}(M) \to 0$$

where λ is an $(-1)^k$-Hermitian form, simple over $\mathbb{Q}\pi$, which induces the form on $K_{k-1}(M)$. Let $e_1,\ldots,e_m$ be a prefered basis for G and $e_1^*,\ldots,e_m^*$ the dual basis.

Write $A_\lambda(e_i) = e_j^* a_{ji}$, $a_{ji} \in \mathbb{Z}\pi$. Then $\phi(e_j^* a_{ji}) = 0$, so if S_j is an embedded $(k-1)$-sphere which represents $\phi(e_j^*)$, $S_j a_{ji}$ is a boundary in M, say $\partial\xi_i = S_j a_{ji}$. Attach handles at the S_i in $M \times I$ to get $(N;M,M')$. We have $\partial\eta_i = -S_i$, where η_i is the core of the handle at S_i. It follows that $A_\lambda(e_i)$ lifts to the class $f_i \in K_k(N)$ represented by $\eta_j a_{ji} + \xi_i$.

Here we are using

$$K_k(N) \to K_k(N,M) \to K_{k-1}(M) \to 0$$

$$G \cong K_k(N,M'),\ G^* \cong K_k(N,M) \cong K^k(N,M').$$

Let $\tilde{\eta}_p$ denote the result of moving η_p a small distance so that η_p and $\tilde{\eta}_p$ are disjoint. Then the image of f_i in G is $e_p b_{pi}$, $b_{pi} = \lambda(\tilde{\eta}_p, \eta_j a_{ji} + \xi_i)$

$$= \lambda(\tilde{\eta}_p, \xi_i)$$

$$= -\lambda(S_p, \xi_i).$$

Let r be an integer with $r\pi = 0$. Then $rS_i = \partial\Sigma_i$ for some k-chain Σ_i and $r\xi_i - \Sigma_j a_{ji}$ is a cycle. Therefore

$$rb_{pi} = r\lambda(\tilde{S}_p, \xi_i)$$

$$= \lambda(\tilde{S}_p, \Sigma_j a_{ji})$$

$$= \sigma_{pj} a_{ji}$$

where $\sigma_{pj} = \lambda(\tilde{S}_p, \Sigma_j)$.

Choose κ_{pj} so that $re_j^* = A_\lambda(e_p \kappa_{pj})$; by the geometric construction, $\lambda_0(\phi(e_j^*), \phi(e_p^*)) = \frac{1}{r}\sigma_{pj}$ and by hypothesis, it also equals $\frac{1}{r}\kappa_{pj}$. Let $\rho_{ij} = \frac{1}{r}(\kappa_{ij} - \sigma_{ij})$, and subject the spheres S_i to simultaneous disjoint regular homotopies with mutual intersections ρ_{ij} to get embedded spheres $\hat{S}_i$.

Do surgery on the $\hat{S}_i$ to get $(\hat{N};M,\hat{M})$. Let $\hat{f}_i \in K_k(\hat{N})$ represent $\hat{\eta}_j a_{ji} + \xi_i$; then the image of $\hat{f}_i$ in $G = K_k(\hat{N},\hat{M})$ is $e_p\hat{b}_{pi}$, where $\hat{b}_{pi} = b_{pi} + \rho_{pj} a_{ji}$

$$= b_{pi} + \frac{1}{r}(\kappa_{pj} - \sigma_{pj})a_{ji}$$

$$= \frac{1}{r}\kappa_{pj} a_{ji}.$$

Thus $A_\lambda(e_p\hat{b}_{pi}) = A_\lambda(\frac{1}{r}e_p\kappa_{pj}a_{ji}) = e_i^*$, and so the diagram

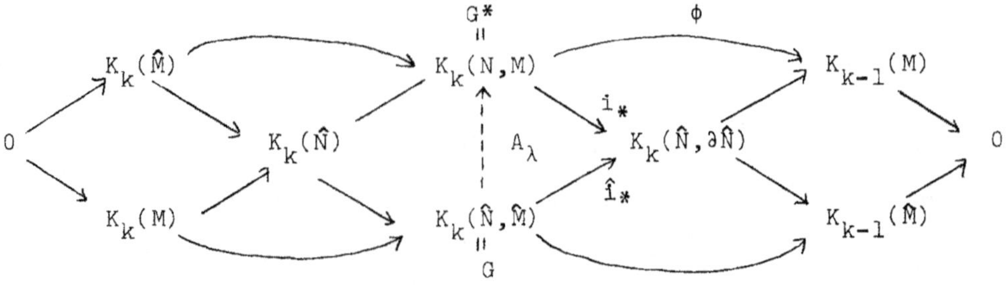

commutes.

Since $i_* \oplus \hat{i}_*$ is onto (by Mayer-Vietoris) and $i_*A_\lambda = \hat{i}_*$, i_* is onto. Thus $K_{k-1}(\hat{M}) = 0$ and $K_k(\hat{M}) = 0$ by duality. Since (a_{ji}) is elementary over $\mathbb{Q}\pi$, as is (κ_{ji}), $f:M \to X$ has torsion in $K(\pi)$. This concludes the proof.

In the non-simple case, this result is due to Wall [H18] and our proof follows his.

We now prove our form of Connolly's realizalization theorem [K6].

<u>Theorem 2. Let</u> (K,λ_0,μ_0) <u>be a simple linking form,</u> X^{2k-2} <u>a manifold with</u> $\pi_1(X) = \pi$, $k \geq 3$. <u>Then there is a normal map</u> $F:(W;\partial_-W,\partial_+W) \to (X\times I;X\times 0 \cup \partial X\times I, X\times 1)$ <u>so that</u>

(i) $F|\partial_-W$ <u>is the identity,</u>

(ii) $F|\partial_+W$ <u>is a homology equivalence over</u> $\mathbb{Z}$ <u>with torsion in</u> $K(\pi)$<u>, and</u>

(iii) <u>the linking form on</u> $K_{k-1}(W)$ <u>coincides with the form on</u> K.

Proof: Let G_1, G_2 be free $\mathbb{Z}\pi$-modules so that we have an exact sequence

$$0 \to G_1 \xrightarrow{f} G_2 \xrightarrow{\phi} K \to 0,$$

where f is a simple isomorphism over $\mathbb{Q}\pi$. Let $e_1, \ldots, e_m$, and $e_1^*, \ldots, e_m^*$ be bases for G_1 and G_2 respectively. Add m trivial handles of index k to XxI to get a manifold N and a normal map $N \to XxI$.

Then $K_{k-1}(\partial_+ N)$ is a free $\mathbb{Z}\pi$-module with generators

$$\begin{aligned} x_i &= i\text{-th sphere } 1xS^{k-1} \\ y_i &= i\text{-th sphere } S^{k-1}x1 \end{aligned} \qquad i = 1, \ldots, m.$$

This gives $K_{k-1}(\partial_+ N)$ the structure of a standard kernel, and so

$$\mu\left(\sum_{i=1}^{m} x_i a_i + y_i b_i\right) = \sum_{i=1}^{m} b_i^* a_i.$$

Here μ denotes the usual self-intersection number.

Write $f(e_i) = e_j^* a_{ji}$ as before and let A be the matrix with a_{ji} in the j-i-th spot if the above equation holds and 0 otherwise. By hypothesis, A is elementary over $Q\pi$. Choose $c_{ij} \in \lambda_0(\phi(e_i^*), \phi(e_j^*))$ and let $C = (c_{ij})$; define B =CA. Then the entries of B are in $\mathbb{Z}\pi$, since for some h,

$$\begin{aligned} b_{ij} &= c_{ih} a_{hj} \\ &= \lambda_0(\phi(e_i^*), \phi(e_h^*) a_{hj}) \bmod \mathbb{Z}\pi \\ &= \lambda_0(\phi(e_i^*), \phi f(e_h)) \\ &= \lambda_0(\phi(e_i^*), 0) \\ &= 0. \end{aligned}$$

Also, $(A^*B)^* = B^*A = A^*C^*A = (-1)^k A^*B$, since $C^* = (-1)^k C$, and the diagonal terms of A^*B are in I_k since

$$\begin{aligned} 0 &= \mu_0(\phi f(e_i)) \\ &= \mu_0(\phi(e_j^* a_{ji})) \\ &= a_{ji}^* c_{jj} a_{ji} \bmod I_k. \end{aligned}$$

It follows that there is a matrix Q over $\mathbf{Z}\pi$ so that

$$A^*B = Q + (-1)^k Q^*.$$

Let $u_i = \sum_{j=1}^{m} (x_j b_{ji} + y_j a_{ji})$ in $K_{k-1}(\partial_+ N)$. Then $\mu(u_i) = 0$, $\lambda(u_i, u_j) = 0$, $i \neq j$, and so we can do surgery on embedded spheres representing the classes u_i. Let W be the trace of the surgeries (including the initial surgeries above). Then

$$\begin{array}{ccccccccc} 0 & \longrightarrow & G_1 & \xrightarrow{\ f\ } & G_2 & \xrightarrow{\ \phi\ } & K & \longrightarrow & 0 \\ & & \downarrow\cong & & \downarrow & & \downarrow\cong & & \\ 0 & \longrightarrow & C_k(W,X) & \xrightarrow{\ \partial\ } & C_{k-1}(W,X) & \xrightarrow{\ p\ } & K_{k-1}(W) & \longrightarrow & 0 \end{array}$$

where the vertical isomorphisms are given by

$G_1 \to C_k(W,X)$, $e_i \mapsto g_i^!(D^k x 0)$ where $g_i^!:(D^k, S^{k-1}) x D^{k-1} \to (N, \partial_+ N)$ extends $g_i: S^{k-1} x D^k \to \partial_+ N$ which represents u_i, and

$G_2 \to C_{k-1}(W,X)$, $e_i^* \mapsto y_i$. The diagram commutes since

$\partial g_i^!(D^k x 0) = y_j a_{ji} = f(e_i)$. Thus we have an isomorphism $K \to K_{k-1}(W)$.

Let $\lambda_0^!$, $\mu_0^!$ denote the linking and self-linking forms on $K_{k-1}(W)$. We must show $\lambda_0^!(p(y_i), p(y_j)) = c_{ij}$. Choose r so that $r\pi = 0$. Then $rG_2 \subset f(G_1)$, so

$$re_j^* = f\Big(\sum_{i=1}^{m} e_i a_{ij}^!\Big)$$

for some $a_{ij}^! \in \mathbf{Z}\pi$. Let $A' = (a_{ij}^!)$; then $AA' = rI$ and

$$\begin{aligned} ry_j &= \sum_{i=1}^{m} g_i^!(D^k x 0) a_{ij}^! \\ &= \sum_{i=1}^{m} g_1(S^{k-1} x 0) a_{ij}^!. \end{aligned}$$

So $\lambda_0'(p(y_i),p(y_j)) = \frac{1}{r}\lambda(y_i, \sum_{h=1}^{m} g_h(S^{k-1}x0)a'_{hj})$

$$= \frac{1}{r}\sum_{h=1}^{m} \lambda(y_i, g_h(S^{k-1}x0)a'_{hj})$$

$$= \frac{1}{r}\sum_{h=1}^{m}\sum_{q=1}^{m}(\lambda(y_i,y_q)a_{qh}+\lambda(y_i,x_q)b_{qh})a'_{hi}$$

$$= \frac{1}{r}\sum_{h=1}^{m} b_{ih}a'_{hj}$$

$$= c_{ij} \bmod \mathbb{Z}\pi, \text{ since } \tfrac{1}{r}BA' = BA'(AA')^{-1} = BA = CA^2$$

and $A^2 = aI$ for some $a \in \mathbb{Z}$. A similar result holds for self-linking numbers. Part(ii) follows immediately from the fact that $\partial: C_k(W,X) \to C_{k-1}(W,X)$ has matrix A. This completes the proof.

Our main theorem is:

<u>Theorem 3</u>. $\operatorname{coker}(\partial_*) \cong \mathcal{L}^s_{2k-1}(\pi,w)$.

Proof: The constructions in Lemma 2 define a map $Lk: \operatorname{coker}(\partial_*) \to \mathcal{L}^s_{2k-1}(\pi,w)$ which is well-defined and injective by Theorem 1 and surjective by Theorem 2. The only thing we need to check is that the form on $K_{k-1}(M)$ is a linking form. Assume M is a triple $(M;\partial_-M,\partial_+M)$; then (M,∂_-M) has a handle decomposition with handles in dimensions k and k-1 only. The sequence

$$0 \to C_k(M,\partial_-M) \to C_{k-1}(M,\partial_-M) \to K_{k-1}(M) \to 0$$

satisfies (ii) of the definiton.

To see that $(K_{k-1}(M),\lambda_0,\mu_0) \oplus (K_{k-1}(M),-\lambda_0,-\mu_0)$ is standard, notice that this is the linking form of $M + (-M)$ and apply Theorem 1.

Corollary 1. For π a finite group, $k > 3$, there is an exact sequence

$$0 \to \mathcal{L}^s_{2k-1}(\pi,w) \to L^{K(\pi)}_{2k-1}(\pi,w) \to L^s_{2k-1}(\pi,w;\mathbb{Q}).$$

Proof: Immediate from Theorem 6.1.2 and Theorem 3.

Corollary 2. As in Corollary 1, there is an exact sequence

$$H^{2k}(\mathbb{Z}/2\mathbb{Z};K(\pi)) \to \ker(L^s_{2k-1}(\pi,w) \to L^s_{2k-1}(\pi,w;\mathbb{Q})) \to$$
$$\to \mathcal{L}^s_{2k-1}(\pi,w) \to H^{2k-1}(\mathbb{Z}/2\mathbb{Z};K(\pi)).$$

Proof: We have the following diagram

$$
\begin{array}{ccccccc}
 & & & & H^{2k}(\mathbb{Z}/2\mathbb{Z};K(\pi)) & & \\
 & & & \swarrow & \downarrow & & \\
0 & \longrightarrow & \ker(i[1]_*) & \longrightarrow & L^s_{2k-1}(\pi,w) & \xrightarrow{\;i[1]_*\;} & L^s_{2k-1}(\pi,w;\mathbb{Q}) \\
 & & \downarrow & & \downarrow & & \downarrow = \\
0 & \longrightarrow & \mathcal{L}^s_{2k-1}(\pi,w) & \longrightarrow & L^{K(\pi)}_{2k-1}(\pi,w) & \longrightarrow & L^s_{2k-1}(\pi,w;\mathbb{Q}) \\
 & & & \searrow & \downarrow & & \\
 & & & & H^{2k-1}(\mathbb{Z}/2\mathbb{Z};K(\pi)) & &
\end{array}
$$

with exact rows and columns. It is easy to construct the maps indicated by the dotted lines and to show that the resulting sequence is exact.

Remark: According to Bass [C2], Theorem 7.5,* if π is finite, then there is an exact sequence

$$0 \to SK_1(\mathbb{Z}\pi) \to K_1(\mathbb{Z}\pi) \to K_1(\mathbb{Q}\pi).$$

It follows that

$$K(\pi) \cong SK_1(\mathbb{Z}\pi)/\pi$$

for π abelian, and this allows us to compute $K(\pi)$ in the general case (theoretically anyway).

* Chapter XI.

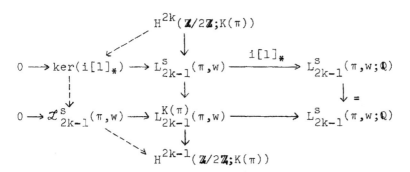

Appendix A. Torsion for n-ads.

In Chapter 5, the torsion of a homology equivalence over a ring of n-ads was (implicitly) defined to be the torsion of the map of underlying spaces. We can generalize this without too much trouble.

Let Λ be a ring of type 2^n, i.e., for each $\alpha \subset \{1,\ldots,n\}$, $\Lambda(\alpha)$ is a ring and homomorphisms $r_{\alpha\beta}:\Lambda(\beta) \to \Lambda(\alpha)$ for $\beta \subset \alpha$. A Λ-module is an object M so that $M(\alpha)$ is a $\Lambda(\alpha)$-module and all maps are compatible. A Λ-homomorphism is defined in the obvious way. The free Λ-module of rank n, Λ^n, is defined by $\Lambda^n(\alpha) = \Lambda(\alpha)^n$.

Define the general linear group $GL(n,\Lambda)$ to be the group of automorphisms of Λ^n; the elementary transformations $E(n,\Lambda)$ are defined as before. Stably, $E(\Lambda) = [GL(\Lambda),GL(\Lambda)]$. Define $K_1(\Lambda) = GL(\Lambda)/E(\Lambda)$.

This definition is equivalent to the definition of K_1 of a (finite) tree, used by Farrell and Wagoner in the theory of infinite simple homotopy types. See [D11].

If π is a group of type 2^n and R is a ring, define $R\pi$ by $(R\pi)(\alpha) = R\pi(\alpha)$. If $R = \mathbb{Z}_p$, let U be the subgroup of $GL(1,\mathbb{Z}_p\pi)$ generated by maps $f:\mathbb{Z}_p\pi \to \mathbb{Z}_p\pi$ of the form $f(\alpha)(\lambda) = rg\lambda$, $\lambda \in \mathbb{Z}_p\pi(\alpha)$, $g \in \pi(\alpha)$, $r \in \dot{\mathbb{Z}}_p$. Define the Whitehead group by

$$Wh(\pi;\mathbb{Z}_p) = K_1(\mathbb{Z}_p\pi)/U.$$

Let $M = (|M|;M_1,\ldots,M_{n-1})$ and $M' = (|M'|;M_1',\ldots,M_{n-1}')$ be manifold n-ads. We say that M and M' are <u>h-cobordant</u> if there is an n-ad $N = (|N|;N_1,\ldots,N_{n-1})$ so that $|N|$ is a manifold with boundary $M \cup N_1 \cup \ldots \cup N_{n-1} \cup M'$, $M \subset |N|$ and $M' \subset |N|$ are homotopy equivalences, and for each α, $N(\alpha)$ is an h-cobordism between $M(\alpha)$ and $M'(\alpha)$.

<u>Theorem 1</u>. <u>If</u> N <u>is an h-cobordism between</u> M <u>and</u> M', <u>then</u> $N \cong M \times I$ <u>if and only if a torsion invariant</u> $\tau(N,M) \in Wh(\pi_1(M))$ <u>vanishes</u>.

Proof: $\tau(N,M)$ is defined to be the torsion of the "chain complex n-ad" $C_*(N,M)(\alpha) = C_*(N(\alpha),M(\alpha))$ in $K_1(\mathbb{Z}\pi_1(M))$, where the contraction δ is constructed by induction. The result follows by the s-cobordism theorem and induction.

Appendix B. Higher L-Theories.

In Chapter 5, we defined surgery obstruction groups by geometric construction. The following question arises: is there an algebraic definition of $L_n(K;R)$? We saw in Chapter 4 that the answer is yes if K is a space. The answer is yes for K a 2-ad and we sketch the construction.

Ignoring torsion and orientation, let $L_m(\pi \to \pi'; \mathbb{Z}_p)$ denote $L_n(K(\pi,1) \to K(\pi',1); \mathbb{Z}_p)$.

Case 1. $n = 2k+1$.

This is done in detail by Wall [M19] in Chapter 7. We consider 4-tuples (G,λ,μ,K) where:

(i) (G,λ,μ) is a $(-1)^k$-Hermitian form over $\mathbb{Z}_p\pi$

(ii) K is a subkernel of $G \otimes_{\mathbb{Z}_p\pi} \mathbb{Z}_p\pi'$.

Define $(G,\lambda,\mu,K) \sim (G',\lambda',\mu',K')$ if there is a kernel H over $\mathbb{Z}_p\pi$ with subkernel S so that

(i) $(G,\lambda,\mu) \oplus H \oplus (G',-\lambda',-\mu') = H_1$ is a kernel with subkernel S_1, and

(ii) any automorphism of $H_1 \otimes_{\mathbb{Z}_p\pi} \mathbb{Z}_p\pi'$ taking

$$S_1 \otimes_{\mathbb{Z}_p\pi} \mathbb{Z}_p\pi' \text{ to } K \oplus (S \otimes_{\mathbb{Z}_p\pi} \mathbb{Z}_p\pi') \oplus K' \text{ is in } EU(\mathbb{Z}_p\pi').$$

Then $L_{2k+1}(\pi \to \pi'; \mathbb{Z}_p)$ is isomorphic to the group of objects (G,λ,μ,K) modulo the equivalence relation $\sim$.

Case 2. $n = 2k$.

This was determined by Sharpe [H14]; see also Wall [H19], Chapters 8 and 17G. Let $Wh_2(\pi';\mathbb{Z}_p) = K_2(\mathbb{Z}_p\pi')/\ker \phi|W$, where ϕ is the natural map $St(\mathbb{Z}_p\pi') \to E(\mathbb{Z}_p\pi')$, and W is the subgroup generated by $x_{ij}^{\lambda} x_{ji}^{-\lambda^{-1}} x_{ij}^{\lambda}$, $\lambda \in T' = \langle \mathbb{Z}_p^{\cdot}, \pi' \rangle$. (see the end of Section 1.7).

Define $St\, U_r(\mathbb{Z}_p\pi')$ to be the group of pairs (P,Q) of $(-1)^k$-symmetric forms over $\mathbb{Z}_p\pi'$ (on a free module of dimension r) modulo the relation $(P,Q) \sim (P',Q')$ if there is an X so that

$$\text{(i)} \quad I + (-1)^k QX = A \in SL(r,\mathbb{Z}_p\pi')$$

$$\text{(ii)} \quad P' - P = X + (-1)^k X^* + X^*QX$$

$$\text{(iii)} \quad Q = AQ'A^*.$$

Define $St\, U(\mathbb{Z}_p\pi') \to EU(\mathbb{Z}_p\pi')$ by

$$(P,Q) \mapsto \begin{pmatrix} I & 0 \\ P & I \end{pmatrix} \begin{pmatrix} 0 & I \\ (-1)^{k+1}I & 0 \end{pmatrix} \begin{pmatrix} I & 0 \\ Q & I \end{pmatrix},$$

and we have a natural map $h: St(\mathbb{Z}_p\pi') \to St\, U(\mathbb{Z}_p\pi')$. Let $\widetilde{St\, U(\mathbb{Z}_p\pi')}' = \widetilde{St\, U(\mathbb{Z}_p\pi')}/h(\ker \phi|W)$, where $\sim$ denotes the associated split group, e.g.,

$$\widetilde{U_r(\mathbb{Z}_p\pi)} \text{ is the set of things } \left(\begin{pmatrix} \alpha & \beta \\ \gamma & \delta \end{pmatrix}, \begin{pmatrix} a \\ b \end{pmatrix} \right),$$

$$\begin{pmatrix} \alpha & \beta \\ \gamma & \delta \end{pmatrix} \in U_r(\mathbb{Z}_p\pi), \quad a + (-1)^{k+1}a^* = \alpha^*\gamma, \quad b + (-1)^{k+1}b^* = \delta^*\beta.$$

T' acts on $\widetilde{\mathrm{St}\ U(\mathbb{Z}_p\pi')'}$ by conjugation and define

$$\widetilde{\mathrm{St}\ U(\pi';\mathbb{Z}_p)} = T' \times_{T'} \widetilde{\mathrm{St}\ U(\mathbb{Z}_p\pi')'}/[\pi',\pi'].$$

Let $P(\pi\to\pi';\mathbb{Z}_p)$ be the pull-back of

$$\begin{array}{ccc} & & \widetilde{U(\mathbb{Z}_p\pi)} \\ & & \downarrow \\ \widetilde{\mathrm{St}\ U(\pi';\mathbb{Z}_p)} & \to & \widetilde{U(\mathbb{Z}_p\pi')}. \end{array}$$

Finally, we have

$$L_{2k}(\pi\to\pi';\mathbb{Z}_p) \cong P(\pi\to\pi';\mathbb{Z}_p)/\langle T\cdot[P,P],\ Wh_2(\pi';\mathbb{Z}_p)\rangle.$$

Considerations like these can be used to give algebraic definitions of surgery groups to solve problems of the form "how to get a homology equivalence over R on $\mathrm{Int}(M)$ and a homology equivalence over R' on ∂M". This is similar to the problems solved by Cappell and Shaneson's relative Γ-groups, [K5]. Miscenko ([K8]) defines L groups (with $2 \in P$) algebraically by an algebraic bordism procedure. The condition $2 \in P$ (which eliminates self-intersection problems) has been removed (at least in some cases) by Connolly.

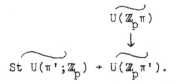

Appendix C. L Groups of Free Abelian Groups.

In this section we compute $L_k(\mathbb{Z}^n;\mathbb{Z}_p)$. The proof is identical to that given in Shaneson [H9] and we refer the reader there for details.

Theorem 1. *There is a split exact sequence* (*for* $n \geq 7$)

$$0 \to L_n{}^s(G,w;\mathbb{Z}_p) \to L_n{}^s(G \times \mathbb{Z},w;\mathbb{Z}_p) \to L^h{}_{n-1}(G,w;\mathbb{Z}_p) \to 0.$$

Proof: The map $L_n{}^s(G,w;\mathbb{Z}_p) \to L_n{}^s(G \times \mathbb{Z},w;\mathbb{Z}_p)$ is the functorial map. Let K be a manifold with $\pi_1(K) = G$. Then every element in $L_n{}^s(G \times \mathbb{Z},w;\mathbb{Z}_p)$ is represented by a normal map $M \overset{\phi}{\to} X \to K \times I \times S^1$ by Theorem 4.3.2.

Assume the composition $X \to K \times I \times S^1 \to S^1$ is a (smooth) fibration f and also $f|\partial X:\partial X \to S^1$. Change ϕ by a homotopy so that the basepoint of S^1 is a regular value of $M \to X \to S^1$. If $(L,\partial L) = f^{-1}(*)$, $(N,\partial N) = (f\phi)^{-1}(*)$, then we get a commutative diagram

$$\begin{array}{ccc} (N,\partial N) & \to & (L,\partial L) \\ \downarrow & & \downarrow \\ (M,\partial M) & \overset{\phi}{\to} & (X,\partial X) \\ & & \downarrow \\ & & S^1. \end{array}$$

Clearly $\phi|N$ is a normal map and by applying the homotopy sequence of a fibration to the fibrations above, $\phi|\partial N:\partial N \to \partial L$

is a homology equivalence over $\mathbb{Z}_p$. Define $L_n^s(G \times \mathbb{Z}, w; \mathbb{Z}_p) \to L^h_{n-1}(G, w; \mathbb{Z}_p)$ by sending ϕ to the class of $\phi|\partial N$. This is a well-defined homomorphism, and the splitting is defined by xS^1, as in Section 5.3. This goes to L_n^s by Kwun and Sczcarba [D16].

The next theorem allows us to apply this.

<u>Theorem 2</u>. <u>Let</u> G <u>be a finitely generated free abelian group</u>. <u>Then</u> $Wh(G; \mathbb{Z}_p) = 0$.

Proof: Assume $G = \mathbb{Z}$. Then $\mathbb{Z}_p[G] \cong \mathbb{Z}_p[t, t^{-1}]$, the ring of Laurant polynomials over $\mathbb{Z}_p$. According to Bass [C2], Chapter XII, Theorem 7.4, there is an exact sequence

$$K_1(\mathbb{Z}_p[t]) \oplus K_1(\mathbb{Z}_p[t^{-1}]) \to K_1(\mathbb{Z}_p[t, t^{-1}]) \to K_0(\mathbb{Z}_p) \to 0$$

But

$$\begin{array}{ccc} K_1(\mathbb{Z}_p[t,t^{-1}]) & & \\ \uparrow & \nwarrow^{\cong} & \\ K_1(\mathbb{Z}_p[t]) & \longleftarrow & K_1(\mathbb{Z}_p) \cong \mathbb{Z}_p^{\bullet} \end{array}$$

and similarly for t^{-1}.

Thus $K_1(\mathbb{Z}_p[t, t^{-1}])/\mathbb{Z}_p^{\bullet} \cong K_0(\mathbb{Z}_p) \cong \mathbb{Z}$, and the result follows. The general case follows by induction and the fact $\mathbb{Z}_p[G \times \mathbb{Z}] \cong \mathbb{Z}_p[G][t, t^{-1}]$.

<u>Corollary 1</u>. $L_n(\mathbb{Z}^k; \mathbb{Z}_p) \cong \bigoplus_{i=0}^{n} L_{n-i}(1; \mathbb{Z}_p)$.

Appendix D. Ambient Surgery and Surgery Leaving a Submanifold Fixed.

In this section we give some results on submanifolds. For simplicity, we will assume $P = \phi$ and $A = 0$ so we can involke the s-cobordism theorem. Also, because of transversality problems with Poincare complexes, only manifolds will be considered. (see Jones [E12] for a discussion of transversality obstructions.) Suitable modifications can be made to get the general case.

Suppose $g:N \to Y$ be a simple homotopy equivalence between manifolds of dimension n. If X is a submanifold of Y, is g homotopic to g_0 so that $g_0|g_0^{-1}X:g_0^{-1}X \to X$ is a simple homotopy equivalence? This is the problem of ambient surgery.

Suppose X has dimension $k \geq 5$ and let E be a closed regular neighborhood of X in Y.

Assume g is transverse to X and put $M = g^{-1}X$; let F be a regular neighborhood of M in N.

Theorem 1. *There are two obstructions* $\sigma(g|M) \in L_k^s(\pi_1 X;\mathbb{Z})$, $\theta \subset L_n^s(\pi_1(Y-X) \to \pi_1 Y;\mathbb{Z})$ *so that* $g \simeq g_0$, *with* $g_0|g_0^{-1}X \to X$ *a simple homotopy equivalence if* $\sigma = 0$ *and* $0 \in \theta$.

Proof: Suppose $\sigma = 0$. Then it follows that the surgery obstruction of $(F,\partial F) \to (E,\partial E)$ vanishes. Let $f:(P,Q) \to (E,\partial E)$ be a cobordism to a simple homotopy equivalence. Form the normal map

$$\psi : N \times I \cup P \to Y \times I$$

defined by $g \times 1 : N \times I \to Y \times I$ and $f : P \to E$. Then the obstruction to doing surgery relative to $N \times 0 \cup R$, where R is the right boundary of P, lies in $L_n{}^s(\pi_1(Y-X) \to \pi_1 Y)$. Let θ be the set of all such obstructions for all cobordisms (P,Q).

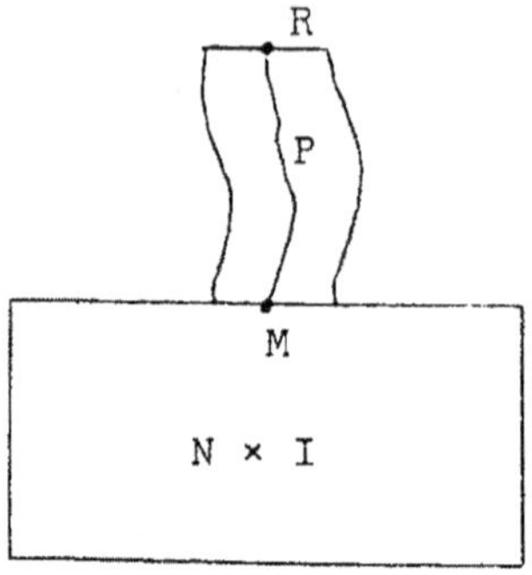

If $0 \in \theta$, then ψ is cobordant to a simple homotopy equivalence $\psi' : (W; \partial_- W, \partial_+ W) \to (Y \times I; Y \times 0, Y \times 1)$, leaving $N \times 0$ and R fixed. By the s-cobordism theorem, $W \cong N \times I$. This gives the result.

This also gives the obstruction to the ambient surgery problem. To see this, suppose $g \simeq g'$ and g' is transverse to X. Let $F : N \times I \to Y$ be a homotopy between g and g'. Assume $F' : N \times I \to Y \times I$ is transverse to $X \times I$, where $F'(x,t) = (F(x,t),t)$. Then $F'^{-1}(X \times I)$ is a cobordism between $g^{-1}(X \times 0)$ and $g'^{-1}(X \times 1)$, and so the obstructions (which are cobordism invariants) are the same. Thus the obstructions are independent of which way we make g transverse to X.

Note by Theorem 5.1.1, $L_n^s(\pi_1(Y-X) \to \pi_1 Y) = 0$ if $\pi_1(Y-X) \xrightarrow{\cong} \pi_1 Y$, e.g., if $n-k \geq 3$.

A more careful analysis has been made by Cappell and Shaneson [K5] in the codimension 2 case and applied to knot theory. The simply connected case is due to Browder [G2]. Se also Lopez de Medrano [J22], sections III.3.2 and VI.3.

A similar procedure, modified by an idea of Wall [A21] can be used to do surgery leaving a submanifold fixed. As before, this is essentially a codimension 0 problem, i.e., we look only at the regular neighborhoods.

First an example. Define a homomorphism $p: L_{n+1}(\mathbb{Z}^n) \to L_{n+1}(\mathbb{Z}^{n-1})$ as follows: by Appendix C, $L_m(\mathbb{Z}^k)$ is a finite direct sum of cyclic groups of order 2 or ∞. Let $e_1,\ldots,e_r$ be a set of generators of $L_{n+1}(\mathbb{Z}^n)$. By Theorem 4.3.2, represent e_i by $\phi_i:(W_i;\partial_- W_i,\partial_+ W_i) \to (T^n \times I; T^n \times 0, T^n \times 1)$ where T^n is the n-torus. Let $p(e_i)$ = the surgery obstruction of

$$W_i \cup_{\partial_- W_i} D^2 \times T^{n-1} \to T^n \times I \cup D^2 \times T^{n-1} \times 0 \simeq D^2 \times T^{n-1}.$$

Since $L_{n+1}(\mathbb{Z}^n) \cong L_{n+1}(\mathbb{Z}^{n-1}) \oplus L_n(\mathbb{Z}^{n-1})$, there exists a non-zero element in $\ker(p)$. This is represented by a map $W \to T^n \times I$ so that $W \cup D^2 \times T^{n-1} \to T^n \times I \cup D^2 \times T^{n-1}$ has surgery obstruction 0. Thus we cannot do surgery leaving $D^2 \times T^{n-1}$ fixed.

Let $\phi : N \to Y$ be a normal map, where N and Y are manifolds of dimension $n \geq 5$. Suppose ϕ is cobordant to a simple homotopy equivalence and that M is submanifold of N of dimension k so that $\phi|M : M \to X$ is a simple homotopy equivalence.

Theorem 2. There is an obstruction $\theta \subset L_n^s(\pi_1(Y-X) \to \pi_1(Y);\mathbb{Z})$ so that if $0 \in \theta$, then ϕ is cobordant to $\phi_+ : N_+ \to Y$ with

(i) ϕ_+ is a simple homotopy equivalence

(ii) $M = \phi_+^{-1}X$

(iii) $\phi_+|M \simeq \phi|M$.

Proof: Let $\psi : Q \to Y$ be a cobordism between ϕ and a simple homotopy equivalence $\phi_+ : N_+ \to Y$. Assume ψ is transverse to X and let $W = \psi^{-1}(X)$. Then W is a cobordism between $\phi|M : M \to X$ and $\phi_+|M_+ : M_+ \to X$, $M_+ = \phi_+^{-1}X$. Let D, E be regular neighborhoods of W, X in Q, Y, respectively.

Then ψ extends to $\hat{\psi} : D \cup N_+ \times I \to E \times I \cup Y \times I$, and $\hat{\psi}$ is a simple homotopy equivalence on $\partial_- D \cup (N_+ \times 1)$. It follows that the obstruction to doing surgery on

$\hat{\psi}$ rel $\partial_- D \cup (N_+ \times 1)$ to get a simple homotopy equivalence lies in $L_n^s(\pi_1(Y-X) \to \pi_1 Y)$. Define θ by considering different cobordisms Q.

If $0 \in \theta$, then $\hat{\psi}$ is cobordant to a simple homotopy equivalence $\psi': V \to E \times I \cup Y \times I$, and assume ψ' is transverse to $E \times 1 = E$. Let $A = (\psi')^{-1}(E \times I)$, $B = (\psi')^{-1}(Y \times I)$, $C = (\psi')^{-1}(E \times 1)$.

Since $\pi_1(E) \cong \pi_1(E \times I)$, $\psi':(A,C) \to (E \times I, E)$ is cobordant to a simple homotopy equivalence $(A',C') \to (E \times I, E)$ by a cobordism $\psi_0:(Z,Z_0) \to (E \times I, E)$. Let $U = V \times I \cup Z$ along $A \times 1$. Then we have a normal map of triads induced by ψ', $\psi_T':(U; V \times 0 \cup A', B \times 1 \cup Z_0) \to ((E \times I \cup Y \times I) \times I;$

$$(E \times I \cup Y \times I) \times 0 \cup (E \times I) \times 1, (Y \times I) \times 1).$$

Now ψ_T' is a simple homotopy equivalence on $V \times 0 \cup A'$, and $\pi_1((Y \times I) \times 1) \to \pi_1((E \times I \cup Y \times I) \times I)$ is an isomorphism, so we can do surgery on ψ_T' to get $(P; P_1, P_2)$. As before, P is an s-cobordism and so $P \cong V \times I$. Thus ψ' is homotopic to a simple homotopy equivalence

$$(V; A', B'; C') \to (E \times I \cup Y \times I; E \times I, Y \times I; E)$$

where $B' = V - (A'-C')$. So A' is an s-cobordism between D and C', and B' is an s-cobordism between $N_+' = \partial_- B'$ and $N_+ \times 1 = \partial_+ B'$.

Let $R = N \times I \cup A' \cup B'$. Then ∂R is the disjoint union of $N \times 0$, $N_+ \times 1$ and $N'' = N \times 1 \cup \partial A' \cup N_+ \times 0$.

We can extend our maps to a normal map $\hat{\phi}:R \to Y$. It follows that $\hat{\phi}$ is a cobordism between $\phi + (-\phi_+)$ and $\phi'':N'' \to Y$. Since ϕ is cobordant to ϕ_+, ϕ'' is null-cobordant; say $\hat{\phi}'':P'' \to Y$ is a cobordism bounding ϕ''. Let $T = R \cup P''$ and extend our maps to $T \to Y$.

Since A' and B' are products, ϕ is cobordant to ϕ_+ leaving M fixed. This concludes the proof.

Appendix E. Homotopy and Homology Spheres.[1]

In this section we describe the calculation of groups of homotopy or homology spheres over $\mathbb{Z}_P$. This is applied to the theory of resolutions of $\mathbb{Z}_P$-homology manifolds.

Let $H = PL$ or DIFF. A closed H-manifold Σ^n is a $\mathbb{Z}_P$-<u>homology sphere</u> if $H_*(\Sigma;\mathbb{Z}_P) \cong H_*(S^n;\mathbb{Z}_P)$. A cobordism $(W;M,M')$ is an <u>H-cobordism</u> over $\mathbb{Z}_P$ if $H_*(W,M;\mathbb{Z}_P) = H_*(W,M';\mathbb{Z}_P) = 0$. We let $\psi_n^P(H)$ denote the group of H-cobordism over $\mathbb{Z}_P$ classes of H $\mathbb{Z}_P$-homology n-spheres (addition given by connected sum).

A $\mathbb{Z}_P$-homology sphere Σ is a $\mathbb{Z}_P$-<u>homotopy sphere</u> if $\pi_1(\Sigma) = 0$. We let $\theta_n^P(H)$ denote the group of h-cobordism over $\mathbb{Z}_P$ classes of H $\mathbb{Z}_P$-homotopy n-spheres. Generalizing the method of [G20], we have

<u>Theorem 1. Let</u> $n \geq 4$ <u>and</u> $2 \notin P$. <u>Then</u> $\theta_n^P(H) \cong \psi_n^P(H)$.

Define a homomorphism $r: L_{4k}(1;\mathbb{Z}_P) \longrightarrow \theta_{4k-1}^P(PL)$ by sending x to the $\mathbb{Z}_P$-homotopy sphere bounding the manifold obtained by plumbing with x (see Section 4.4). Surgery arguments show that r defines injection

$r': L_{4k}(1;\mathbb{Z}_P)/L_{4k}(1) \longrightarrow \theta_{4k-1}^P(PL)$.

[1]Added May 1976

Theorem 2. For $n \geq 4$,

$$\theta_n^P(PL) \cong \begin{cases} H_n & n \not\equiv 3 \bmod (4) \\ H_n \oplus L_{4k}(1;\mathbb{Z}_P)/L_{4k}(1) \oplus \bigoplus_{\pi(k)-1} \mathbb{Z}_P/\mathbb{Z} & n=4k-1 \end{cases}$$

where H_n is a finite P-torsion group and $\pi(k)$ is the number of partitions of k.

For the proofs of Theorems 1 and 2, see [K13].

In the smooth case, we have the results of Barge, Lannes, Latour and Vogel [K14]: A left inverse to the map r' above can be defined in the smooth homology sphere case; we let $\tilde{\psi}_{4k-1}^P$ be the kernel of this map, $\tilde{\psi}_n^P = \psi_n^P(DIFF)$ if $n \not\equiv 3 \bmod (4)$. Then for $n \geq 4$,

$$\tilde{\psi}_n^P = \tilde{\psi}_n^{\phi} \otimes \mathbb{Z}_P \oplus \tilde{\psi}_n^{Q} \otimes \mathbb{Z}_{(P)} .$$

Thus, the calculations of $\tilde{\psi}_n$ [G20], [F1] and $\tilde{\psi}_n^{\mathbb{Q}}$ [K14] are enough to characterize $\tilde{\psi}_n^P$.

In the case $2 \not\in P$, it follows from [K2] that

$$bP_{4k}^P/bP_{4k} \cong a_P\mathbb{Z}/8\mathbb{Z} \oplus \bigoplus_{p\in P} W_p^P , \text{ where } bP_{4k}^P \subset \theta_{4k-1}^P(DIFF) \text{ is the}$$

the group of spheres which bound $\mathbb{Z}_K$-parallelizable manifolds.

One application of this is to the study of $\mathbb{Z}_P$-homology manifolds. Let $f:M \longrightarrow N$ be a surjective PL-map between polyhedra. We say f is a $\tilde{\mathbb{Z}}_p$-*resolution* if $\tilde{H}_*(f^{-1}(x);\mathbb{Z}_p) = 0$ for each $x \in N$.

Theorem 3. (Sullivan) Let N^n *be a* $\tilde{\mathbb{Z}}_p$-*homology manifold. Then there is a* $\tilde{\mathbb{Z}}_p$-*resolution* $f:M^n \longrightarrow N$ *to a PL-manifold* M *if and only if obstructions in* $H^k(N;\psi^P_{k-1}(PL))$ *vanish.*

References

A. Background in Topology and Algebra.

1. Atiyah, M. and MacDonald, I., Introduction to Commutative Algebra. Addison-Wesley, 1969.

2. Bredon, G., Sheaf Theory. McGraw-Hill, 1967.

3. Browder, W., Liulevicius, A., and Peterson, F., Cobordism theories. Ann. of Math. 84 (1966) 91-101.

4. Cerf, J., Sur les Diffeomorphismes de la sphere de Dimension trois ($\Gamma_4 = 0$). Springer Lecture Notes #53.

5. Conner, P. and Floyd, E., Differentiable Periodic Maps. Springer-Verlag, 1964.

6. Gabriel, P. and Zisman, M., Calculus of Fractions and Homotopy Theory. Springer-Verlag, 1967.

7. Hirzebruch, F., Topological Methods in Algebraic Geometry. Springer-Verlag, 1966.

8. Hudson, J., Piecewise-Linear Topology. Benjamin, 1969.

9. Kan, D., Abstract Homotopy Theory III. Proc. Nat. Acad. Sci. 42 (1956) 419-21.

10. Lam, T., Algebraic Theory of Quadradic Forms, Benjamin, 1973.

11. Milnor, J., The geometric realization of a semi-simplicial set. Ann. of Math. 65 (1957) 357-62.

12. ________, and Husemoller, D., Lectures on Quadratic Forms. Springer-Verlag, 1973.

13. Namioka, I., Maps of pairs in homotopy theory. Proc. London Math. Soc. 34 (1962) 725-38.

14. Rohlin, V., A new result in the theory of 4-dimensional manifolds. Doklady 8 (1952) 221-4.

15. Spanier, E., Algebraic Topology. McGraw-Hill, 1966.

16. Stong, R., Notes on Cobordism Theory. Princeton, 1968.

17. Thom, R., Quelques proprietes globales des varietes differentiables. Comm. Math. Helv. 28 (1954) 17-86.

19. Wall, C.T.C., Classification Problems in Differential Topology I,II. Topology 2 (1963) 253-72.

20. ___________, Determination of the cobordism ring. Ann. of Math. 72 (1960) 292-311.

21. ___________, Cobordism of pairs. Comm. Math. Helv. 35 (1961) 136-45.

22. Whitney, H., The self intersection of a smooth n-manifold in 2n-space. Ann. of Math. 45 (1944) 220-46.

23. Williamson, J., Cobordism of combinatorial manifolds. Ann. of Math. 83 (1966) 1-33.

See also B12, C2.

24. Serre, J., A Course in Arithmetic, Springer, 1970.

25. Puppe, D., Homotopiemenge und ihre induzierten Abbildungen, I, Math. Z. 69 (1958) 299-344.

26. MacLane, S., Homology, Springer, 1967.

B. Immersions and Embeddings.

1. Haefliger, A., Lissage des Immersions I. Topology 3 (1967) 221-39.

2. ______________, Lectures on the theorem of Gromov. in Proc. Liverpool Singularities Symposium II, Springer Lecture Notes.

3. ______________, and Poenaru, V., La classification des immersions combinatoires. Publ. Math. I.H.E.S. 23 (1964) 75-91.

4. Hirsch, M., Immersions of manifolds. Trans. A.M.S. 93 (1959) 242-76.

5. Hudson, J. and Zeeman, E., On regular neighborhoods. Proc. London Math. Soc. 14 (1964) 719-45.

6. Kuiper, N. and Lashof, R., Microbundles and bundles I: Elementaty theory. Inv. Math. 1 (1966) 1-17.

7. Mazur, B., Relative neighborhoods and the theorems of Smale. Ann. of Math. 77 (1963) 232-49.

8. Milnor, J., Microbundles I. Topology 3 (Supp.)(1964) 53-80.

9. Rourke, C., and Sanderson, B., Block Bundles I,II,III. Ann. of Math. 87 (1968) 1-28,256-78,431-83.

10. ______________________________, Homotopy theory of Δ-sets. preprint.

11. ______________________________, Some results on topological neighborhoods. Bull. A.M.S. 76 (1970) 1070.

12. ______________________________, Introduction to Piecewise Linear Topology. Springer-Verlag. 1972.

See also F2.

C. Algebraic K-Theory.

1. Alpern, R., Dennis, K., and Stein, M., The non-triviality of $SK_1(Z\pi)$. in Proc. of the Conference on Orders, Group Rings and Related Topics. Springer Lecture Notes #353.

2. Bass, H., Algebraic K-Theory. Benjamin, 1969.

3. ________, K-theory and stable algebra. Publ. Math. I.H.E.S. 22 (1964) 5-60.

4. ________, Topics in Algebraic K-Theory. Tata Institute, Bombay, 1967.

5. Gelfand, I. and Miscenko, A., Quadradic forms over commutative group rings and K-theory. Functional Analysis 3 (1965) 28-33. (in Russian).

6. Heller, A., Some exact sequences in algebraic K-theory. Topology 3 (1965) 389-408.

7. Higman, G., Units of group rings. Proc. London Math. Soc. 46 (1940) 231-48.

8. Milnor, J., Introduction to Algebraic K-Theory. Princeton, 1971.

9. Siebenmann, L., The Obstruction to Finding a Boundary for an Open Manifold of Dimension greater than five. Doctoral thesis. Princeton, 1965.

10. Wall, C.T.C., Finiteness conditions for CW complexes, I,II. Ann. of Math. 81 (1965) 56-69; Proc. Roy. Soc. A 295 (1966) 129-139.

11. Swan, R., Algebraic K-Theory, Springer Lecture Notes 1968.

D. Whitehead Torsion and h-Cobordisms.

1. Algebraic K-Theory and its Geometric Applications. Springer Lecture Notes #108.

2. Barden, D., Structure of Manifolds. Thesis, Cambridge University 1963.

3. ________, h-cobordisms between 4-manifolds. Notes, Cambridge University, 1964.

4. Bass, H., Heller, A., and Swan, R., The Whitehead group of a polynomial extension. Publ. Math. I.H.E.S. 22 (1964) 61-79.

5. Chapman, T., Compact Hilbert cube manifolds and the invariance of Whitehead torsion. Bull. A.M.S. 79 (1973) 52-6.

6. Cohen, M., A Course in Simple Homotopy Theory. Springer-Verlag 1973.

7. DeRham, G., Kervaire, M., and Maumary, S., Torsion et Type Simple d'Homotopy. Springer Lecture Notes #48.

8. Edwards, R., The topological invariance of simple homotopy type for polyhedra. preprint.

9. Farrell, F., and Hsiang W., A formula for $K_1R_a[T]$. in Proc. Symp. in Pure Math. 17 (Categorical Algebra). Amer. Math. Soc. 1970.

10. ________________, h-cobordant manifolds are not necessarily homeomorphic. Bull. A.M.S. 73 (1967) 72-77.

11. ________, and Wagoner, J., Infinite matrices in algebraic K-theory and topology. Comm. Math. Helv. 47 (1972) 474-501.

12. ________________, Algebraic torsion for infinite simple homotopy types. ibid. 502-13.

13. Golo, V., Realization of Whitehead torsion and discriminants of bilinear forms. Sov. Math. Doklady 9 (1968) 1532-4.

14. Hsiang, W., A splitting theorem and the Kunneth formula in algebraic K-theory. in D1.

15. Kervaire, M., La theorem de Barden-Mazur-Stallings. Comm. Math. Helv. 40 (1965) 31-42. also in D7.

16. Kwun, K., Szczarba, R., Product and sum theorems for Whitehead torsion. Ann. of Math. 82 (1965) 183-190.

17. Lawson. T., Inertial h-cobordisms with finite cyclic fundamental group. Proc. A.M.S. 44 (1974) 492-96.

18. Mazur, B., Differential topology from the point of view of simple homotopy theory. Publ. Math. I.H.E.S. 15 (1963) 5-93.

19. Milnor, J., Lectures on the h-Cobordism Theorem. Princeton, 1965.

20. ________, Whitehead Torsion. Bull. A.M.S. 72 (1966) 358-426.

21. Siebenmann, L., Infinite simple homotopy types. Indag. Math. 32 (1970) 479-95.

22. Smale, S., Generalized Poincare's conjecture in dimensions greater than four. Ann. of Math. 74 (1961) 391-406.

23. ________, On the structure of manifolds. Amer. J. Math. 84 (1962) 387-99.

E. Poincare Duality.

1. Allday, C., and Skjelbred, T., The Borel formula and the topological splitting principal for torus actions on a Poincare duality space. Ann. of Math. 100 (1974) 322-25.

2. Bredon, G., Fixed point sets of actions on Poincare duality spaces. Topology 12 (1973) 159-75.

3. Browder, W., Poincare spaces, their normal fibrations and surgery. Inv. Math. 17 (1972) 191-202.

4. ___________, The Kervaire invariant, products and Poincare transversality. Topology 12 (1973) 145-58.

5. ___________, and Brumfiel, G., A note on cobordism of Poincare duality spaces. Bull. A.M.S. 77 (1971) 160.

6. Chang, T. and Skjelbred, T., Group actions on Poincare duality spaces. Bull. A.M.S. 78 (1972) 1024-26.

7. Cohen, J., A note on Poincare 2-complexes. Bull. A.M.S. 78 (1972) 763.

8. Hodgson, J., The Whitney approach for Poincare complex embeddings. Proc. A.M.S. 35 (1972) 263-68.

9. ___________, Poincare complex thickenings and concordance obstructions. Bull. A.M.S. 76 (1970) 1039-43.

10. ___________, General position in the Poincare duality category. Inv. Math. 24 (1974) 311-34.

11. ___________, Subcomplexes of Poincare complexes. Bull. A.M.S. 80 (1974) 1146-50.

12. Jones, L., Patch Spaces: A geometric representation for Poincare spaces. Ann. of Math. 97 (1973) 306-43.

13. Lashof, R., Poincare duality and cobordism. Trans. A.M.S. 109 (1963) 257-77.

14. Levitt, N., The structure of Poincare duality spaces. Topology 7 (1968) 369-88.

15. ___________, Generalized Thom spectra and transversality for spherical fibrations. Bull. A.M.S. 76 (1970) 727-31.

16. ___________, Poincare duality cobordism. Ann. of Math. 96 (1972) 211-44.

17. ___________, and Morgan, J., Transversality structures and p.l. structures on spherical fibrations. Bull. A.M.S. 78 (1972) 1064.

18. Quinn, F., Surgery on normal spaces. Bull. A.M.S. 78 (1972) 262-67.

19. Spivak, M., Spaces satisfying Poincare duality. Topology 6 (1967) 77-102.

20. Stasheff, J., A classification theorem for fiber spaces. Topology 2 (1963) 239-46.

21. Wall, C.T.C., Poincare complexes I. Ann. of Math. 86 (1967) 213-45.

F. Surgery.

1. Kervaire, M. and Milnor, J., Groups of homotopy spheres. Ann. of Math. 77 (1963) 504-37.

2. Lees, J., Immersions and surgeries of topological manifolds. Bull. A.M.S. 75 (1969) 529-34.

3. Milnor, J., A proceedure for killing the homotopy groups of differentiable manifolds. in Proc. Symp. in Pure Math. 3 (Diff. Geo.) Amer. Math. Soc. 1961 39-55.

4. Wall, C.T.C., Killing the middle homotopy groups of odd dimensional manifolds. Trans. A.M.S. 103 (1962) 421-33.

5. Wallace, A., Modifications and cobounding manifolds.I,II, III. Can. J. Math. 12 (1960), 503-28; J. Math. Mech. 10 (1961) 773-809; ibid 11 (1961) 971-90.

G. Simply Connected Surgery and Applications.

1. Bernstein, I., A proof of the vanishing of the simply-connected surgery obstruction in the odd-dimensional case. preprint, Cornell University 1969.

2. Browder, W., Embedding 1-connected manifolds. Bull. A.M.S. 72 (1966) 225-31, 736.

3. __________, Surgery and the theory of differentiable transformation groups. in Proc. Conference on Transformation Groups.(New Orleans, 1967). Springer-Verlag 1968.

4. __________, The Kervaire invariant of framed manifolds and its generalization. Ann. of Math. 90 (1969) 157-86.

5. __________, Surgery on Simply Connected Manifolds. Springer-Verlag 1972.

6. Brumfiel, G. and Morgan, J., Quadratic functions, the index modulo 8, and a Z/4-Hirzebruch formula. Topology 12 (1973) 105-22.

7. Milgram, J., Surgery with coefficients. Ann. of Math. 100 (1974) 194-248.

8. Morgan, J. and Sullivan, D., The transversality characteristic class and linking cycles in surgery theory. Ann. of Math. 99 (1974) 463-544.

9. Novikov, S., Diffeomorphisms of simply connected manifolds. Doklady 3 540-43.

10. Orlik, P., Seminar Notes on Simply Connected Surgery. Notes, IAS 1968.

11. Rourke, C., The Hauptvermutung According to Sullivan I,II. Notes, IAS 1967.

12. __________, and Sullivan, D., On the Kervaire Obstruction. preprint, University of Warwick 1968.

13. Sullivan, D., Triangulating Homotopy Equivalences. Thesis, Princeton University, 1965.

14. __________, On the Hauptvermutung for manifolds. Bull. A.M.S. 73 (1967) 598-600.

15. __________, Geometric Topology: Localization, Periodicity, and Galois Symmetry. Notes, M.I.T. 1970.

16. __________, Geometric periodicity and the invariants of manifolds. in Manifolds-Amsterdam 1970. Springer Lecture Notes #197.

17. __________, Genetics of homotopy theory and the Adams conjecture. Ann. of Math. 100 (1974) 1-79..

18. Wall, C.T.C., An extension of results of Novikov and Browder. Amer. J. of Math. 88 (1966) 20-32.

19. __________, Non-additivity of the signature. Inv. Math. 7 (1969) 269-74.

20. Kervaire, M., Smooth homology manifolds and their fundamental groups, Trans. Amer. Math. Soc. 144 (1969), 67-72.

H. Surgery Obstruction Groups.

1. Browder, W., Manifolds and Homotopy theory. in Lecture Notes #197 (see G16).

2. ______, Manifolds with π_1 = Z. Bull. A.M.S. 72 (1966) 238-44.

3. ______, Homotopy type of differentiable manifolds. in Colloq. on Algebraic Topology. notes, Aarhus, 1962.

4. ______, and Hirsch, M., Surgery on PL-manifolds and applications. Bull. A.M.S. 73 (1967) 242-45.

5. ______, and Quinn, F., A surgery theory for G-manifolds and stratified sets. preprint.

6. Lees, J., The surgery groups of C.T.C. Wall. Adv. in Math. 11 (1973) 113-56.

7. Maumary, S., Proper surgery groups and Wall-Novikov groups. in I1.

8. Quinn, F., A Geometric Formulation of Surgery. Thesis, Princeton, 1970. see also article in Georgia conf. on Topology of Manifolds.

9. Shaneson, J. Wall's surgery obstruction groups for ZxG. Ann. of Math. 90 (1969) 296-334.

10. ______, Product formulas for $L_n(\pi)$. Bull. A.M.S 76 (1970) 787.

11. ______, Hermitian K-theory in topology. in I1.

12. ______, Some problems in hermitian K-theory. in I1.

13. ______, Non-simply connected surgery and some results in low-dimensional topology. Comm. Math. Helv. 45 (1970) 333-52.

14. Sharpe, R., Surgery on Compact Manifolds: The Bounded Even Dimensional Case. Thesis, Yale 1970. also in Ann. of Math. 98 (1973) 187-209.

15. ______, Surgery and unitary K_2. in I1.

16. Taylor, L., Surgery on Paracompact Manifolds. Thesis, Berkeley, 1971.

17. Wagoner, J., Smooth and piecewise linear surgery. Bull. A.M.S. 73 (1967) 72-77.

18. Wall, C.T.C., Surgery of non-simply-connected manifolds. Ann. of Math. 84 (1966) 217-76.

19. ___________, Surgery on Compact Manifolds. Acad. Press 1970.

20. Williamson, R., Surgery in MxN with $\pi_1 M \neq 1$. Bull. A.M.S. 75 (1969) 582-85.

I. Algebraic L-Theory.

1. Algebraic K-Theory III: Hermitian K-Theory and Geometric Applications. Battelle Inst. Conf. 1972. Springer Lecture Notes #343.

2. Bak, A., The Stable Structure of Quadratic Modules. Thesis, Columbia University 1969.

3. _______, On modules with quadratic forms. in D1.

4. _______, The computation of surgery groups of odd torsion groups. Bull. A.M.S. 80 (1974) 1113-16.

5. _______, and Scharlau, W., Grothendieck and Witt groups of orders of finite groups. Inv. Math. 23 (1974) 207-40.

6. Bass, H., Unitary algebraic K-theory. in I1.

7. _______, L_3 of finite abelian groups. Ann. of Math. 79 (1974) 118-53.

8. Cappell, S., Mayer-Vietoris sequences in hermitian

K-theory. in Il.

9. __________, Unitary Nilpotent Groups and Hermitian K-Theory. Bull. A.M.S. 80 (1974) 1117-22.

10. Frohlich, A. and McEvett, A., Forms over rings with involution. J. Algebra 12 (1969) 79-104.

11. Karoubi, M., Some problems and conjectures in algebraic K-theory. in Il.

12. __________, Periodicite de la K-theorie hermitianne. CR Acad. Sci. Paris t 273 (1971). also in Il.

13. Lee, R., Computation of Wall Groups. Topology 10 (1971).

14. Novikov, S., The algebraic construction and properties of Hermitian analogues of K-theory for rings with involution from the point of view of Hamiltonian formalism. Some applications to differential topology and the theory of characteristic classes. I,II. Izv. Akad. Nauk. SSSR ser. mat. 34 (1970) 253-88, 475-500.

15. Ranicki, A., Algebraic L-Theory I,II,III. Proc. London Math. Soc. 27 (1973) 101-25, 126-58, and in Il.

16. Wall, C.T.C., Quadratic forms on finite groups and related topics. Topology 2 (1963) 281-98.

17. __________, On the axiomatic foundations of the theory of hermitian forms. Proc. Camb. Phil. Soc. 67 (1970) 243-50.

18. __________, On the Classification of Hermitian Forms.

I. Rings of algebraic integers. Comp. Math.

II. Semisimple rings. Inv. Math. 18 (1972) 119-41.

III. Complete semilocal rings. Inv. Math. 19 (1973) 59-71.

IV. Adele rings. Inv. Math. 23 (1974) 241-60.

V. Global rings. ibid. 261-88.

19. __________, Foundations of algebraic L-theory. in Il.

20. __________, Some L-groups of finite groups. Bull. A.M.S. 79 (1973) 526-30.

21. Taylor, L., Surgery groups and inner automorphisms. in Il.

J. Applications of Surgery.

1. Browder, W., Embedding smooth manifolds. In Proc. I.C.M. (Moscow) 1966.

2. ________, Structures on MxR. Proc. Camb. Phil. Soc. 61 (1965) 337-45.

3. ________, Free Z_p-actions on homotopy spheres. in Georgia Conference on Topology.

4. ________, and Levine, J., Fibering manifolds over a circle. Comm. Math. Helv. 40 (1966) 153-60.

5. ________, and Livesay, G., Fixed point free involutions on homotopy spheres. Bull. A.M.S. 73 (1967) 242-45.

6. ________, Petrie, T., and Wall, C.T.C., The classification of free actions of cyclic groups of odd order on homotopy spheres. Bull. A.M.S. 77 (1971) 455.

7. Cappell, S., A splitting theorem for manifolds and surgery groups. Bull. A.M.S. 77 (1971) 281-86.

8. ________, Splitting obstructions for Hermitian forms and manifolds with $Z_2 \subset \pi_1$. Bull. A.M.S. 79 (1973) 909-13.

9. ________ and Shaneson, J., Surgery on 4-manifolds and applications. Comm. Math. Helv. 46 (1971) 500-28.

10. ________________, Submanifolds, group actions and knots I,II. Bull. A.M.S. 78 (1972) 1045.

11. Casson, A., Fibrations over spheres. Topology 6 (1967) 489-99.

12. Farrell, F., The Obstruction to Fibering a Manifold over a Circle. Thesis, Yale 1967. also in Bull. A.M.S. 73 (1967) 737-40.

13. Hsiang, W. and Shaneson, J., Fake tori, the annulus conjecture and the conjectures of Kirby. Proc. Nat. Acad. Sci. 62 (1969) 687-91. see also article in Topology of Manifolds.

14. ________ and Wall, C.T.C.,;On homotopy tori II. Bull. London Math. Soc. 1 (1969) 341-42.

15. Kirby, R. and Siebenmann, L., On the triangulation of manifolds and the Hauptvermutung. Bull. A.M.S. 75 (1969) 742-49. see also article in Manifolds-

Amsterdam 1970.

16. Lashof, R. and Rothenberg, M., On the Hauptvermutung, triangulation of manifolds, and h-cobordism. Bull. A.M.S. 72 (1966) 1040-43.

17. ______________________________, Triangulation of manifolds I,II. Bull. A.M.S. 75 (1969) 750-57.

18. Lee, R., Splitting a manifold into two parts. preprint IAS 1968.

19. ______, Semicharacteristic classes. Topology 12 (1973) 183-199.

20. ______ and Orlik, P., On a codimension 1 embedding problem. preprint, IAS 1969.

21. Levitt, N., Fiberings and manifolds and transversality. Bull. A.M.S. 79 (1973) 377-81.

22. Lopez de Medrano, S., Involutions on Manifolds. Springer-Verlag 1970.

23. Novikov, S., Homotopy Equivalent Smooth Manifolds I. Translations A.M.S. 48 (1965) 271-396.

24. __________, Manifolds with free abelian fundamental groups and their applications (Pontrjagin classes, smoothness, multidimensional knots). Translations A.M.S. 71 (1968) 1-42.

25. Petrie, T., The Atiyah-Singer invariant, the Wall groups $L_n(\pi,1)$ and the function te^x+1/te^x-1. Ann. of Math. 92 (1970) 174-87.

26. _________, Induction in equivariant K-theory and geometric applications. in I1.

27. Quinn, F., $B_{(TOP_n)^{\sim}}$ and the surgery obstruction. Bull. A.M.S. 77 (1971) 596-600.

28. Wall, C.T.C., Free piecewise linear involutions on spheres. Bull. A.M.S. 74 (1968)554-58.

29. ___________, On homotopy tori and the annulus theorem. Bull. London Math. Soc. 1 (1969) 95-97.

30. ___________, The topological space-form problem. in Topology of Manifolds.

31. Farrell, F. and Hsiang, W., Manifolds with $\pi_1 = G x_\alpha Z$. Amer. J. Math. XCV (1973) 813-48.

K. Surgery with Coefficients.

1. Agoston, M., The reducibility of Thom complexes and surgery on maps of degree d. Bull. A.M.S. 77 (1971) 106.

2. Alexander, J., Hamrick, G., and Vick, J., Involutions on homotopy spheres. Inv. Math. 24 (1974) 35-50.

3. Anderson, G., Surgery with Coefficients and Invariant Problems in Surgery. Thesis, University of Michigan 1974.

4. Cappell, S., Groups of singular hermitian forms. in Il.

5. ___________ and Shaneson, J., The codimension two placement problem and homology equivalent manifolds. Ann. of Math. 99 (1974) 277-348.

6. Connolly, F., Linking numbers and surgery. Topology 12 (1973) 389-410.

7. Geist, R., Semicharacteristic Detection of Obstructions to Rational Homotopy Equivalences. Thesis, Notre Dame, 1974. see also Notices A.M.S. 21, 1974.

8. Miscenko, A., Homotopy invariants of non-simply connected manifolds. I. Rational invariants. Izv. Akad. Nauk SSSR ser mat 34 (1970) 501-14.

9. Pardon, W., Thesis, Princeton University, 1974.

10. Passman, D., and Petrie, T., Surgery with coeffients in a field. Ann. of Math. 95 (1972) 385-405.

11. Alexander, J., Hamrick, G., and Vick, J., Cobordisms of manifolds with odd order normal bundles. Inv. Math. 24 (1974) 83-94.

12. Anderson, G., Computation of the surgery obstruction groups $L_{4k}(1;\mathbb{Z}_p)$, (to appear, Pacific Math. J.).

13. ___________, Groups of Λ-homology spheres, (preprint).

14. Barge, J., Lannes, J., Latour, F., and Vogel, P., Λ-spheres, Ann. Sci. Norm. Sup., 1974.

15. Barge, J., Structures differentiables sur les types d'homotopic rationnelle simplement connexes.

INDEX

adjoint 60
Arf invariant 77
Atiyah, M. and
MacDonald, I. 24
Ambient surgery 129

Bass, H. 21, 121, 128
Bernstein, I. 74
block bundle 7
Bredon, G. 40
Browder, W. 46, 78, 131

Cappell, S. and
Shaneson, J. 28, 84, 104,
126, 131
classifying space for
surgery 95
cobordism extension
property 104
Cohen, M. 37
colocalization 27
conjugate closed subgroup 41
Connolly, F. 116, 126

deformation 32
degree 1 41, 86
dimension (of a Poincare
complex) 39
dual (of a module) 1
duality theorem 43
Δ-map 4
Δ-set 4

elementary matrix 20
elementary P-collapse,
expansion 32

Farrell, F. and Wagoner, J. 122
free and based module 21
free module 1
formal deformation 32
fundamental class 39

Gabriel, P. and Zisman, M. 5
Gauss, C. 75
general linear group 20, 122
geometric realization 5
group ring 2
groupoid of type 2^n 92

Haefliger, A. and Poenaru, V. 12
handle 54
handle subtraction 82
Hasse-Minkowski invariant 76
h-cobordism 38
Hermitian form 60
Hirsch, M. 12
homology equivalence 3
homology intersection pairing 14
homology manifold 40
homology type 3
Hudson, J. 38

immersion classification theorem 12
infinite simple homotopy type 122
intersection numbers 14

Jones, L. 129

Kan, D. 5
kernel 60
Kervaire, M. 38
Kervâire, M. and Milnor, J. 78
Kervaire manifold 78
Kirby, R. and Siebermann, L. 54
Kwan, K. and Szczarba, R. 128

Lam, T. 75
Lees, J. 12
linking forms, group of 112
, simple 112
, standard simple 112
local homotopy (homology) 25
local n-sphere 25
localization, algebraic 23
, geometric 24
, relative 26
Lopez de Medrano, S. 131

manifold n-ad 86
microbundle 10
Milnor Poincare complex (manifold) 7
Milnor, J. 5, 7, 10, 17, 22, 28,
29, 54
and Husemoller, D. 76
Miscenko, A. 126
Morse theory 54

Symbol Index

A_λ 60

$(BG)_P$ 47

BH, BA 5

$CH_n^A(\pi,w;R)$ 101

$C^{(n)}$ 85

$C_*(X)$ 2

$C_*(f)$, f a map 42

$C_*(f)$, f a map of pairs 42, 83

$\partial_i, \partial, \delta_i$ 85

$D_-^k, D_+^k, \partial_- D^k, \partial_+ D^k$ 33

$E(u,\Lambda)$, $E(\Lambda)$ 20

$EU_k^A(u,\Lambda)$ 63

Ex^∞ 5

$GL(u,\Lambda)$, $GL(\Lambda)$ 20,122

G_P/H 48

$G_q(R)$ 7

$H^n(\mathbb{Z}/2\mathbb{Z};\ )$ 109

H_q, $\tilde{H}_q$ for H=TOP, PL, DIFF 6

I_k 16, 60

$Imm(M,N)$ 11

$K_0(\Lambda)$ 19

$K_1(\Lambda)$ 21, 122

$\bar{K}_1(\Lambda)$ 21

$K_2(\Lambda)$ 23

$K_i(\ )$, $K^i(\ )$ 3

$K(\pi)$ 110

$K(\pi,1)$, π a groupoid of type 2^n 92

$L^A{}_{2k}(\Lambda)$ 62

$L^A{}_{2k+1}(\Lambda)$ 64

$L_m(\Pi)$, $L_m'(\Pi)$ 87

$L_m^A(\bar{K};R)$ 92

$\mathcal{L}_m(g;\Pi,\Pi')$ 103

$\mathcal{L}^s{}_{2k-1}(\pi,w)$ 112

$L_m^h(\bar{K};R)$, $L_m^s(\bar{K};R)$ 92

$\Lambda^\cdot$ 1

$\mathcal{N}_P^H(X)$ 80

$\pi_r(\phi)$, ϕ a map 55

$\pi_r(\psi)$, ψ a map of pairs 82

$\pi_r(X)$, X an n-ad 86

$\Pi(P)$ 24

$Q_m(\Pi)$ 87

$R\pi$ 2

$R(TM,TN)$ 11

$\mathcal{R}_m(\Pi)$ 87

$\mathcal{R}_m(\Pi',\Pi)$ 103

$SK_1(\Lambda)$ 21

$s_{n,k}$ 85

$S^{-1}\Lambda$ 23

$St(n,\Lambda)$, $St(\Lambda)$ 22

σ 63

$*$ (on $R\pi$) 2

$*$ (on $K_1(\Lambda)$) 22

TM 10

$\widetilde{TM}$ 8

$U_k^A(n,\Lambda)$ 63

$Wh(\pi;R)$ 28

$Wh_2(\pi;\mathbb{Z}_p)$ 125

Ω_n^e 94

$\Omega_m(\Pi)$ 87

$\Omega_m^e(\bar{K})$ 87

xN 96

$[X]$ 39

$[X,\partial X]$ 41

X_p 24

X^p 27

$\mathbb{Z}_p$, $\mathbb{Z}_{(p)}$ 24

n-ad 85
normal cobordism 51
invariant 51
map 50, 57

Pardon, W. 104
periodicity isomorphism 96, 110
plumbing theorem 72
Poincare complex 39
n-ad 86
pair 41
preferred base 21
principal H-bundle 5
A-fibration 5
projective module 1
π-π theorem 83

Quinn, F. 94, 95

realization theorem 72
ring with involution 1
Rothenberg, M. 105
Rourke, C. and Sanderson, B. 5, 7, 17, 54

s-basis 1
s-cobordism 38
s-cobordism theorem 38, 123
self-dual 41
self-intersection number 16
Serre, J. 76
s-free 1
Shaneson, J. 96, 105, 127
Sharpe, R. 125
Siebenmann, L. 19
signature 75
simple chain complex 22
equivalence 22
homology equivalence 29
homology type 36
Poincare complex 39
spherical fibration (over a ring) 10
Spanier, E. 22, 30, 97
Spivak, M. 10, 45
Spivak normal fibration 45
split group 125
stable basis 1
stably free 1
standard plane 60
Steinberg group 23
Stong, R. 94

subkernel 61
Sullivan, D. 25, 47, 50, 80
surgery 54
hypothesis 64
leaving a sub-manifold fixed 131
obstruction theorem 64, 93
rel the boundary 55
with coefficients 57
Swan, R. 21

tangent block bundle 8
microbundle 10
Thom space 45
torsion for n-ads 122
of a chain complex 22
of a Hermitian form 60
of a map 29
of a Poincare complex 39
trace 54
transfer 39

unitary Steinberg group 125

Wall, C.T.C. 19, 41, 47, 58, 64, 80, 84, 95, 96, 116, 124, 125, 131
Wall group 62, 64, 87, 92, 124, 126
Whitehead, J.H.C. 37
Whitehead group of a group 28
of a ring 21
, secondary 125
Whitehead lemma 20
Whitney lemma 16
Williamson, R. 96